ALGEBRA

the text of this book is printed on 100% recycled paper

ALGEBRA

GERALD E. MOORE

Revised

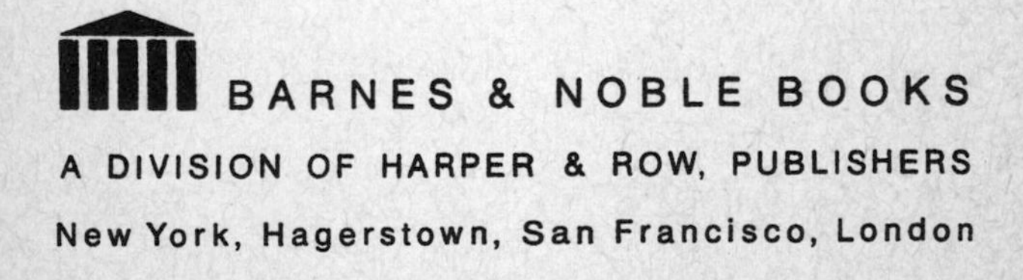
BARNES & NOBLE BOOKS
A DIVISION OF HARPER & ROW, PUBLISHERS
New York, Hagerstown, San Francisco, London

L. C. Catalogue Card Number : 51-2943

ISBN: 0-06-460038-6

Manufactured in the United States of America

78 79 80 12 11 10 9

Note to Students

This outline has been written particularly for readers who have had at least one year of high school algebra. However, it is complete in itself and may be read with profit by those who are beginning the subject. It is presented with the hope that students may find it a useful supplement to their textbook and work in class. It may also serve the purpose of giving any reader a knowledge of algebra, without the labor of solving many problems.

It includes the usual topics studied in a first college course in algebra. Each new term or concept is defined at the first time it enters into the discussion. The various phases of algebra are illustrated by several hundred solved problems and figures. For some phases of the subject, particularly for the Theory of Equations and for Determinants, the theorems are stated in general terminology and proved only for special cases. It is hoped that this treatment will give the reader a clear insight into the nature of the arguments used in proving the general theorems.

Two chapters have been devoted to the theory of problem solving. Chapter XI, *The Nature of Problem Solving*, attempts to give the student a better understanding of the use of axioms, rules of logic, analysis, and tests of conclusions, than is found in most textbooks. Chapter XII, *Solutions of Typical Problems*, gives a detailed discussion of many of the types of problems which usually cause students considerable trouble. This chapter also illustrates some of the practical applications which may be made of a knowledge of algebra.

Several copies of final examinations (with answers) have been included. These should be an aid to students in preparing for their own hour quizzes and final examinations. Careful use of them should give the student confidence in evaluating for himself his mastery of algebra.

G. E. M.

ABOUT THE AUTHOR

The late Gerald E. Moore received his B.S. degree from Hedding College and the degrees of M.S. and Ph.D. from the University of Illinois. He held teaching positions in Hanover College, Illinois State Normal University, and the University of Illinois. For several years he was in charge of the undergraduate courses in mathematics at the University of Illinois, where he was Associate Dean of the College of Arts and Sciences from 1943 to 1960. Professor Moore was a member of the editorial committee (with Brenke, Torrance, and Cell) which published *Engineering Problems Illustrating Mathematics* (McGraw-Hill Book Company, Inc.) as a project of the Mathematics Division of the Society for the Promotion of Engineering Education. He was coauthor (with Crathorne) of the college text, *Brief Trigonometry* (Henry Holt and Company).

Table of Contents

Chapter I

SOME FACTS FROM ELEMENTARY ALGEBRA

1. Introduction.

Before taking up a systematic treatment of the usual topics which constitute a course in college algebra, it will be helpful for us to discuss some of the elementary concepts which the student has already met in his high school algebra course.

2. Some Classes of Numbers.

We shall begin by recalling that numbers as we have already met them have been divided into three classes:

a) Integers: whole numbers such as 1, 2, 7, 43, 967, -32, etc.
b) Rational fractions: such as $\frac{1}{2}$, $\frac{3}{4}$, $-\frac{2}{3}$, $\frac{7}{8}$, $-\frac{9}{11}$, etc.
c) Irrational numbers: such as $\sqrt{2}$, $\sqrt{3}$, etc.

We shall give definitions of these three types of numbers (all of which are real numbers), as occasion arises. For the present we shall confine our attention to integers and rational fractions.

3. The Fundamental Operations.

The fundamental operations are:

a) Addition
b) Subtraction
c) Multiplication
d) Division
e) Raising quantities to powers
f) Extraction of roots of quantities.

We shall presently discuss each of these operations in detail with respect to its applications to various classes of numbers.

4. The Unit 1.

In order to have a system of enumeration, we shall designate 1 as the unit quantity. This unit has the following properties:

a) 1 added to itself gives a new quantity 2, etc.
b) 1 added to any integer N gives the next higher integer $N + 1$.
c) 1 multiplied by any number N gives N.
d) Any number N divided by 1 gives N.

5. Negative Numbers.

These numbers enter in various ways into our scheme of enumeration. For example, if the temperature drops low enough we begin to talk about "degrees below zero"; even the daily newspapers carry accounts of temperatures of -5, -17, etc., to indicate that below-zero weather is being experienced. We therefore assume that the reader is already familiar with the concept of a negative number. These numbers also arise from the operation of subtracting a given positive quantity from a smaller positive quantity.

6. The Number Zero.

Another number which needs special consideration is zero. It is neither positive nor negative. We usually think of it as separating the positive from the negative numbers.

If we delineate numbers as positions along a line, we have the usual scheme such as:

−5 −4 −3 −2 −1 0 1 2 3 4 5

This scheme gives rise to the idea of order among numbers, and we say that -4 is less than -2, -1 is less than zero or any of the positive numbers.

The number zero, designated 0, has some distinct properties as follows:

a) Zero added to or subtracted from any number N gives N.
b) Zero times any number N gives 0.
c) The division of any number N by zero cannot be defined. (Some texts express this fact by saying that division by zero is not a permissible operation.) We shall discuss the details of this situation in Sec. 55.
d) Zero divided by any number N (not zero) is zero.

7. Addition.

Whereas in arithmetic we have used all of the fundamental operations as they apply to numerical values, we now wish to study these operations as extended to apply to more general numbers. Part of our work will be the representation of numbers by literal values. Such literal values will represent either fixed values (or constants) or variable numbers (or variables).

By the sum of two quantities 2a and 3a, which is 5a, we mean:

$$2a + 3a = 5a.$$

Also $(2x^2 + 5ax + 7) + (x^2 + ax + 3) = 3x^2 + 6ax + 10.$

Furthermore, $6 + 3 = 3 + 6$, or more generally $a + b = b + a$. This latter expression means that the sum of two or more quantities is the same in whatever order they are added. This is known as the **commutative law** for addition.

Finally, $a + b + c = (a + b) + c = a + (b + c)$. This means that the sum of several quantities is the same regardless of the manner in which the partial sums are grouped; *i.e.*, $3 + 2 + 4 = (3 + 2) + 4 = 3 + (2 + 4)$; or: $9 = 5 + 4 = 3 + 6$, etc. This is known as the **associative law** for addition.

8. Subtraction.

In arithmetic we learned that $7 - 3 = 4$. In the same sense we have for algebra, $3x - 2x = x$. Again, $4x^2 + x + 11$ minus $x^2 + 5x + 3$ yields $3x^2 - 4x + 8$. This may be obtained by writing:

$$\begin{array}{rr} & 4x^2 + x + 11 \\ \text{minus} & x^2 + 5x + 3 \\ \hline & 3x^2 - 4x + 8 \end{array}$$

or: $(4x^2 + x + 11) - (x^2 + 5x + 3) = 3x^2 - 4x + 8,$

and one may think of this as being arrived at by changing the signs in the subtrahend and proceeding as under the operation of addition.

Subtraction is, then, the process by which we find **b** say from the relationship $a + b = x$ when we are given **a** and **x**. This value **b** is expressed algebraically $b = x - a$. Hence we may look upon subtraction as the inverse of addition. If **a** is greater than **x** (expressed symbolically $a > x$), then **b** is negative. If

$a = x$, then $b = 0$. And if **a** is less than **x** (expressed $a < x$), **b** is positive.

Since $4 - 7 = -3$ is not the same as $7 - 4 = +3$, we see that the commutative law does not hold for subtraction.

The associative law, however, does hold for subtraction. Before going further, we must discuss the use of parentheses.

9. Symbols of Grouping.

The signs of grouping ordinarily used are the **parentheses** ( ), **brackets** [ ], **braces** { }, and **vinculum** ——. These symbols are used to bind together certain terms, and to show the order of sequence of the various fundamental operations. Their uses can be best shown by examples. There are a few rules which govern their use, and these we now state:

Rule 1.

Any of the symbols of grouping may be removed from an expression enclosed therein without altering the expression, provided the grouping symbol is preceded by a + sign.

Rule 2.

Conversely: Any expression may be enclosed by symbols of grouping preceded by a + sign, without altering the signs of any of the terms so enclosed.

Rule 3.

If any expression is enclosed by a symbol of grouping preceded by a − sign, the symbol may be removed provided the sign of every term so enclosed is changed.

Rule 4.

Conversely: Any expression may be enclosed by a symbol of grouping preceded by a − sign, provided the sign of each term so enclosed is changed.

Illustration 1.

Simplify the following by removing the symbols of grouping and combining like terms:

$$2x + 3[x - \{2a - (a + 4b)\} + 2(x - 2b)].$$

By removing the innermost symbol of grouping at each step, we have the following equivalent expressions:

$$
\begin{aligned}
&2x + 3\,[x - \{2a - a - 4b\} + 2(x - 2b)] \\
&= 2x + 3\,[x - 2a + a + 4b + 2x - 4b] \\
&= 2x + 3\,[3x - a] \\
&= 2x + 9x - 3a \\
&= 11x - 3a.
\end{aligned}
$$

An equivalent statement for the original expression could be written using the vinculum instead of the parenthesis, as follows:

$$2x + 3[x - \{2a - \overline{a + 4b}\} + 2 \cdot \overline{x - 2b}].$$

Usually the vinculum is not used until the other symbols have been exhausted.

Illustration 2.

Enclose the last three terms of $2x - 3y + 4b - ax - y$ in parentheses preceded by the minus sign.

$$2x - 3y + 4b - ax - y = 2x - 3y - (-4b + ax + y).$$

10. Multiplication.

By the product of two expressions **a** and **b,** we mean $a \times b$ and express this product by a number **c** such that $a \times b = c$.

Another symbol which is used for multiplication is shown by the following symbolic notation:

$$a \times b = a \cdot b = ab = c.$$

Thus, 3×4 means $4 + 4 + 4$, which equals 12. Also 4×3 means $3 + 3 + 3 + 3$, which equals 12. Since $3 \times 4 = 4 \times 3 = 12$ we have our first indication of a principle or law which holds for multiplication.

We say that $a \cdot b = b \cdot a$ and mean by this that the product is the same whatever the order of multiplication. This is known as the **commutative law** for multiplication.

Extending this idea we have,

$$a \cdot b \cdot c = a(b \cdot c) = (a \cdot b)c = (a \cdot c)b, \text{ etc.}$$

As a numerical example:

$$3 \cdot 2 \cdot 4 = 3(2 \cdot 4) = (3 \cdot 2)4 = (3 \cdot 4)2 = 24.$$

This is known as the **associative law** for multiplication.

Another principle of multiplication is illustrated as follows:

$$a(b + c + d) = ab + ac + ad.$$

This is known as the **distributive law.**

If a number **a** is multiplied by itself we write $a \cdot a = a^2$; also, $a \cdot a \cdot a = a^3$; the index 2 or 3, etc., indicating the number of times the quantity **a** is used as a factor.

Example 1: Thus a product like $2ab^2 \cdot 3a^2b^3 = 6a^3b^5$.

Example 2: Also,
$$2ab^2 \cdot 2a^2b \cdot 3ac = 2 \cdot 2 \cdot 3 \cdot a^4b^3c = 2^2 \cdot 3 \cdot a^4b^3c = 4 \cdot 3a^4b^3c = 12a^4b^3c.$$

Example 3: $2a(x^2 - 5ax + bc) = 2ax^2 - 10a^2x + 2abc$.

11. Division.

The quotient obtained by dividing any number **a** by a number **b** (where **b** is not equal to zero) is written $a \div b$ or a/b.

To divide any number **a** by any number **b** means to find a number **x** such that $bx = a$. Division is therefore the process by which this number is found, and is indicated $a/b = x$.

It follows also, since $\frac{12}{3}$ is not equal to $\frac{3}{12}$, that the commutative law does not hold for division.

12. Rule of Signs for Multiplication and Division.

a) The product of quantities of like signs is a positive quantity.

Examples: $3 \cdot 4 = 12$; $(-3)(-4) = +12$; $(-a)(-b) = +ab$.

b) The product of quantities of unlike signs is negative.

Examples: $(-3)(+4) = -12$; $(3)(-4) = -12$; $(-a)(b) = -ab$.

c) The quotient of quantities of like signs is positive.

Examples: $\frac{+12}{+3} = +4$; $\frac{-12}{-3} = +4$; $\frac{(-a)}{(-b)} = +\frac{a}{b}$.

d) The quotient of quantities of unlike signs is negative.

Examples: $\frac{-12}{+3} = -4$; $\frac{-a}{+b} = -\frac{a}{b}$; $\frac{+a}{-b} = -\frac{a}{b}$.

e) The continued product of an even number of negative quantities is a positive quantity.

f) The continued product of an odd number of negative quantities is a negative quantity.

These above facts are stated here as rules. A detailed study of their validity, based on a postulatory basis, is a topic which properly belongs to an advanced course in algebra. Most first courses do not attempt to verify these facts.

13. Products and Quotients of Multinomials.

An expression which contains but a single term is called a **monomial.** Most algebraic expressions contain more than one term, and such expressions are called **multinomials.**

The method of obtaining products and quotients of multinomials will now be illustrated.

Illustration 1.

Find the product obtained by multiplying

$$2x^3 + 7x^2y - 4 \text{ by } x^2 - 2xy.$$

Such a product is found by multiplying each term of the first expression by each term of the second and adding the results. Thus,

$$\begin{array}{lllll}
2x^3 & + 7x^2y & - 4 & & \\
 & x^2 & - 2xy & & \\
\hline
2x^5 & + 7x^4y & - 4x^2 & & \\
 & - 4x^4y & & - 14x^3y^2 & + 8xy \\
\hline
2x^5 & + 3x^4y & - 4x^2 & - 14x^3y^2 & + 8xy
\end{array}$$

Terms of the same degree, such as $+7x^4y$ and $-4x^4y$, are placed in the same column.

Illustration 2.

Find the quotient obtained by dividing

$$8x^5 - 4x^4y + 6x^3y^2 + 4x^3y - 2x^2y^2 + 3xy^3 \text{ by } 2x^2 + y.$$

Here the dividend is already arranged in descending powers of x beginning with $8x^5$, and had it not been arranged in this order, the first step would be to so arrange it. The divisor is likewise arranged before actual division is begun.

$$
\begin{array}{l|llllll|l}
2x^2 + y & 8x^5 & -4x^4y & +6x^3y^2 & +4x^3y & -2x^2y^2 & +3xy^3 & 4x^3 - 2x^2y + 3xy^2 \\
 & 8x^5 & & & +4x^3y & & & \\ \hline
 & & -4x^4y & +6x^3y^2 & & -2x^2y^2 & +3xy^3 & \\
 & & -4x^4y & & & -2x^2y^2 & & \\ \hline
 & & & +6x^3y^2 & & & +3xy^3 & \\
 & & & 6x^3y^2 & & & +3xy^3 & \\ \hline
\end{array}
$$

The result in the quotient is obtained as three distinct terms, namely $4x^3 - 2x^2y + 3xy^2$. The first term of this quotient is obtained as follows: The first term of the divisor is $2x^2$ and is contained in the first term of the dividend $4x^3$ times. The complete divisor is now multiplied by $4x^3$, yielding $8x^5 + 4x^3y$, and these terms are placed directly below similar terms in the dividend before subtracting. The next step is obtained by noting that $2x^2$ is contained in the first term of the partial dividend $-2x^2y$ times and proceeding as before. This process is continued until the division is completed. In the example above there was no final remainder, so that the division was exact. In case there is a final remainder, it is divided by the divisor and added to the quotient. For example, if 38 is divided by 5, we have:

$$
\begin{array}{l|l}
5 & 38 \\ \hline
 & 7 \text{ and a remainder of } 3.
\end{array}
$$

The quotient in this case is $7\frac{3}{5}$.

14. Axioms of Equality.

As in arithmetic and geometry we need to make use of the following axioms:

1. If equals are added to equals, the sums are equal.
2. If equals are subtracted from equals, the differences are equal.
3. If equals are multiplied by equals, the products are equal.
4. If equals are divided by equals, the quotients are equal (provided the divisor is not zero).
5. Expressions which are equal to the same quantity are equal to each other.
6. The same powers of equals are equal.
7. The same roots of equals are equal.

The student should already be familiar with these axioms and should realize that the application of them must be adhered to strictly in manipulating with algebraic expressions. The writer has found frequently that many difficulties and mistakes encountered by students in mathematics are due to the failure to apply some one or more of the above axioms. Attention will be called from time to time to their application in particular instances. Study them carefully!

15. Definitions.

Equation. An **equation** is a statement of equality between two expressions.

Classes of Equations. There are two general classes of equations:

a) **Identical equations** are those which reduce to the same value for both members, regardless of the values assigned to the letters or symbols.

Illustrations:

$$(a + b)^2 = a^2 + 2ab + b^2.$$
$$(x - 3)(x - 2) = x^2 - 5x + 6.$$

b) **Conditional equations** are those which are true for a restricted number of values of the letters or symbols used.

Illustrations:

$2x - 3 = x + 4$ is true **only** if $x = 7$.
$x^2 + x - 12 = 0$ is true **only** if $x = 3$ or $x = -4$.
$2x + 3y = 8$ is true if $x = 1$ and $y = 2$;

as well as an indefinitely large number of pairs of values which may be obtained by choosing a value for x and computing the corresponding value of y.

Equations are made up of **terms,** which are set off from one another by + or − signs. Thus in $ax^3 - 4xy^2 + 7x^2y^3 + 5x - 2by^2 = 0$, there are five terms. The letters **a, b,** and the numerical values are constants, called the **coefficients** of the terms. The values x and y are called the **unknowns** or **variables** of the equation.

Degree of a Term. This is defined as the sum of the exponents of the variables of the term. Thus ax^3 is of degree 3; $4xy^2$ is of degree 1 in x, degree 2 in y, and the total degree of the term is $1 + 2 = 3$; $7x^2y^3$ is of degree 2 in x, degree 3 in y, and of total degree $2 + 3 = 5$.

Degree of an Equation. The degree of the highest degreed term which occurs is taken to be the degree of the equation. In the illustration above, the equation is of degree 5 because this is the highest degree term which occurs.

Equations may be classified as to degree:

Equations of the **first** degree are called **linear,**
Equations of the **second** degree are called **quadratic,**
Equations of the **third** degree are called **cubic,**
Equations of the **fourth** degree are called **quartic,** etc.

The value or values which satisfy an equation are called **solutions** or **roots** of the equation.

CHAPTER II

SPECIAL PRODUCTS AND FACTORING

16. Some Simple Products.

There are certain fundamental products which play such an important part in what is to follow that we now enumerate them.

Each of these depends for its proof upon the actual multiplication.

A) $(x + a)(x + a) = (x + a)^2 = x^2 + 2ax + a^2.$

The second part of relation (A) [that is, $(x + a)^2$] follows at once from the definition of a product of a quantity by itself. The last part is obtained by actual multiplication, thus:

$$\begin{array}{l} x + a \\ \underline{x + a \qquad\qquad} \\ x^2 + ax \\ \underline{\qquad\quad ax + a^2} \\ x^2 + 2ax + a^2 \end{array}.$$

The next few of these products follow, and the reader is urged to verify each of them by performing the multiplication.

B) $(x - a)(x - a) = (x - a)^2 = x^2 - 2ax + a^2.$

C) $(x - a)(x + a) = (x + a)(x - a) = x^2 - a^2.$

D) $(x + a)(x + b) = x^2 + ax + bx + ab$
$= x^2 + (a + b)x + ab.$

E) $(x + a)(x - b) = x^2 + ax - bx - ab$
$= x^2 + (a - b)x - ab.$

F) $(cx + a)(dx + b) = cdx^2 + adx + bcx + ab$
$= cdx^2 + (ad + bc)x + ab.$

17. Applications.

The relationships given in Sec. 16 are useful in forming products, and one considers them as rules for writing the product without performing the actual multiplication. In this sense they play the same role in algebra that the multiplication table does in arithmetic. For example, as soon as one learns the arithmetic multiplication table, the product of $5 \cdot 9$ is written as 45 without adding to find that $9 + 9 + 9 + 9 + 9 = 45$.

Illustration 1.

Write the product of $(x + 3)$ by $(x + 3)$. In this problem the type is (A) of Sec. 16, with $a = 3$. Hence, $(x + 3)(x + 3) = (x + 3)^2 = x^2 + 2 \cdot 3x + 3^2 = x^2 + 6x + 9$.

Illustration 2.

Write the product of $(x - 5)(x + 5)$. This is type (C), with $a = 5$. Thus $(x - 5)(x + 5) = x^2 - 25$.

Illustration 3.

Write the product of $(x + 5)(x + 2)$. This is type (D), with $a = 5, b = 2$.

$$(x + 5)(x + 2) = x^2 + (5 + 2)x + 5 \cdot 2 = x^2 + 7x + 10.$$

Illustration 4.

Write the product of $(x + 5)(x - 3)$. This is type (E). Hence:

$$(x + 5)(x - 3) = x^2 + (5 - 3)x - 5 \cdot 3 = x^2 + 2x - 15.$$

Illustration 5.

Write the product of $(2x + 7)(3x + 5)$. This is type (F), with $c = 2$, $a = 7$, $d = 3$, $b = 5$. Hence:

$$\begin{aligned}(2x + 7)(3x + 5) &= 2 \cdot 3x^2 + (7 \cdot 3 + 5 \cdot 2)x + 7 \cdot 5 \\ &= 6x^2 + (21 + 10)x + 35 = 6x^2 + 31x + 35.\end{aligned}$$

Illustration 6.

Below are listed several examples to illustrate common types.

$$(x - 6)(x - 6) = (x - 6)^2 = x^2 - 12x + 36.$$
$$(x + 2)(x - 6) = x^2 + (2 - 6)x - 12 = x^2 - 4x - 12.$$
$$(x - 2)(x - 3) = x^2 - (2 + 3)x + (-3)(-2) = x^2 - 5x + 6.$$

The special products may also be stated in words. For instance $(a + b)^2 = a^2 + 2ab + b^2$ may be written:
The square of a binomial is equal to the square of the first term plus twice the product of the first and second terms plus the square of the second term.
The type $(a - b)(a + b) = a^2 - b^2$ can be written:

The product of a difference and sum of two quantities is equal to the difference of the squares of the quantities. The reader will find it beneficial to state the special products in words instead of formulas.

18. Other Special Products.

Further types which the student should verify by actual multiplication and then memorize are:

G) $(x + a)^3 = (x + a)(x + a)(x + a) = x^3 + 3ax^2 + 3a^2x + a^3$.

H) $(x - a)^3 = x^3 - 3ax^2 + 3a^2x - a^3$.

I) $(x + a)(x^2 - ax + a^2) = x^3 + a^3$.

J) $(x - a)(x^2 + ax + a^2) = x^3 - a^3$.

K) $(a + b + c)^2 = a^2 + b^2 + c^2 + 2ab + 2ac + 2bc$.

L) $(a - b - c)^2 = a^2 + b^2 + c^2 - 2ab - 2ac + 2bc$.

The two types (K) and (L) are easily seen to be extensions of type (A). Note that in (K) the right-hand expression consists of the squares of each of the three quantities plus twice all the products formed by taking two terms at a time. (That is, twice a and b, twice a and c, and twice b and c, thus exhausting all the types of products taken two at a time.)

Illustration 1.

Write the value of $(x + 2)^3$. This is type (G) with $a = 2$.

Hence:

$$(x + 2)^3 = x^3 + 3(2)x^2 + 3(2^2)x + 2^3 = x^3 + 6x^2 + 12x + 8.$$

Illustration 2.

Write the value of $(2y - 4)^3$. This is type (H) with $x = 2y$ and $a = 4$. Hence:

$$\begin{aligned}(2y - 4)^3 &= (2y)^3 - 3(4)(2y)^2 + 3(4)^2(2y) - (4)^3 \\ &= 8y^3 - (12)4y^2 + (48)(2y) - 64 \\ &= 8y^3 - 48y^2 + 96y - 64.\end{aligned}$$

Illustration 3.

Write the value for $(x - y + 3)^2$. This is type (L).

$$\begin{aligned}(x - y + 3)^2 &= x^2 + y^2 + 3^2 - 2x \cdot y + 2 \cdot x \cdot 3 - 2 \cdot y \cdot 3 \\ &= x^2 + y^2 + 9 - 2xy + 6x - 6y.\end{aligned}$$

Illustration 4.

Write the product of $(x - 2)(x^2 + 2x + 4)$. This is type (J) with $a = 2$. Hence we may write the product immediately knowing that it consists of the difference of the cubes of x and a. Thus:

$$(x - 2)(x^2 + 2x + 4) = x^3 - 2^3 = x^3 - 8.$$

19. Factoring.

The special products have an important use in resolving a given algebraic expression into its component parts, or factors. Their use can best be shown by examples.

Illustration 1.

Find two expressions whose product is $x^2 + ax - bx - ab$. By comparison with type (E) we see that here we have given the right hand member. Hence we can write the factors at once, obtaining: $x^2 + ax - bx - ab = (x + a)(x - b)$.

Another approach to this same problem would be to note that the first two terms contain a factor **x**, and the last two terms contain a factor **b**. Hence we may write:

$$x^2 + ax - bx - ab = x(x + a) - b(x + a) = (x + a)(x - b).$$

Illustration 2.

Find the factors of $x^2 + 7x + 12$. By comparison with (D), $x^2 + (a + b)x + ab$, we see that we must look for two numbers **a** and **b** such that their product is 12 and their sum is 7.

By inspection we see that $a = 4$, $b = 3$, or $a = 3$, $b = 4$ satisfy these conditions. Hence:

$$x^2 + 7x + 12 = (x + 4)(x + 3).$$

Illustration 3.

Factor $x^2 - x - 12$.

In this example we must find two numbers **a** and **b** such that their sum is -1 and their product is -12. Hence:

$$x^2 - x - 12 = (x - 4)(x + 3).$$

Illustration 4.

Factor $x^2 - 13x + 12$.

In this case $a + b = -13$, $a \cdot b = 12$. Hence:

$$x^2 - 13x + 12 = (x - 12)(x - 1).$$

Illustration 5.

Factor $x^2 + 6x + 9$.

This expression is a perfect square. One recognizes a perfect square in the following way. From (A), Sec. 16, $(x + a)^2 = x^2 + 2ax + a^2$. An examination of the right hand member shows that when a binomial is squared, the coefficient of x is equal to twice the square root of the third term a^2. The square root of a^2 is a, and a is 3 in our problem. Hence:

$$x^2 + 6x + 9 = (x + 3)^2.$$

Note 1. The reader should note that an expression such as $x^2 + 10x + 9$ is not a perfect square. For applying the test indicated in this illustration, $a^2 = 9$, $a = 3$, and $2a = 2 \cdot 3 = 6$. The coefficient of x is 10 instead of 6, and thus $x^2 + 10x + 9$ is of type (D) and equals $(x + 9)(x + 1)$.

Illustration 6.

Write the factors of $x^2 - 49$.

This is type (C) with $a^2 = 49$ and $a = 7$. Hence we may write:

$$x^2 - 49 = (x - 7)(x + 7).$$

Illustration 7.

Write the factors of $x^3 + 27$.

This is type (I) with $a^3 = 27$ and $a = 3$. Hence we may write:

$$x^3 + 27 = (x + 3)(x^2 - 3x + 9).$$

Note 2. Later, in Chapter IX, Sec. 68, we show that this second factor has no real linear factors.

Illustration 8.

Write four factors of $y^6 - 64$.

This is type (C) if we think of it as $(y^3)^2 - (8)^2$ where

$x = y^3$ and $a = 8$, and we can now apply our rule to the factoring of the difference of two squares. Consequently,

$$y^6 - 64 = (y^3)^2 - (8)^2 = (y^3 - 8)(y^3 + 8).$$

But each of these factors can be factored again by types (J) and (I) respectively. Therefore:

$$y^6 - 64 = (y - 2)(y^2 + 2y + 4)(y + 2)(y^2 - 2y + 4).$$

Another method of factoring $y^6 - 64$ would be to consider it as $(y^2)^3 - 4^3$ and then apply type (J) with $x = y^2$ and $a = 4$. Thus:

$$\begin{aligned}(y^2)^3 - 4^3 &= (y^2 - 4)(y^4 + 4y^2 + 16)\\ &= (y - 2)(y + 2)(y^2 + 2y + 4)(y^2 - 2y + 4).\end{aligned}$$

The process for obtaining these last two factors from $y^4 + 4y^2 + 16$ will be discussed in the next illustration.

Illustration 9.

Find two factors of $y^4 + 4y^2 + 16$.

This does not come under any of the types which we have discussed but may easily be brought under type (C) in the following manner. By inspection, we note that $y^4 + 8y^2 + 16$ is a perfect square, namely $(y^2 + 4)^2$. In order to obtain this from the given expression we note that the term $4y^2$ must be increased to $8y^2$ by the addition of $4y^2$. But in order to leave the value of the given expression unchanged we must then diminish the same by $4y^2$. Thus by both adding and subtracting $4y^2$, we obtain:

$$\begin{aligned}y^4 + 4y^2 + 16 &= y^4 + 8y^2 + 16 - 4y^2 = (y^4 + 8y^2 + 16) - 4y^2\\ &= (y^2 + 4)^2 - (2y)^2.\end{aligned}$$

(This is now the difference of two squares.)

Factoring according to type (C):

$$(y^2 + 4)^2 - (2y)^2 = [y^2 + 4 - 2y][y^2 + 4 + 2y] = (y^2 - 2y + 4)(y^2 + 2y + 4).$$

Illustration 10.

Find two factors of $x^4 + 4y^4$.

This can be done by the use of the same ideas which were applied in Illustration 9. Note how one discovers a clue to the amount to be added and subtracted.

The square root of x^4 is x^2, the square root of $4y^4$ is $2y^2$. Twice the product of x^2 by $2y^2$ is $2(x^2)(2y^2) = 4x^2y^2$. Therefore by adding and subtracting $4x^2y^2$, we have:

$$x^4 + 4y^4 = x^4 + 4x^2y^2 + 4y^4 - 4x^2y^2 = (x^2 + 2y^2)^2 - (2xy)^2,$$

which now is the difference of two squares.

Hence: $x^4 + 4y^4 = (x^2 + 2y^2 - 2xy)(x^2 + 2y^2 + 2xy)$.

Illustration 11.

Factor $x^2 - y^2 + 2by - b^2$.

By grouping this and enclosing the last three terms in parentheses preceded by a minus sign, we have:

$$x^2 - (y^2 - 2by + b^2) = x^2 - (y - b)^2,$$

which is of type (C). Hence:

$x^2 - y^2 + 2by - b^2 = [x - (y - b)][x + (y - b)] =$
$(x - y + b)(x + y - b)$.

Illustration 12.

Factoring by grouping terms.
Factor $x^2y - 3x^2z + 2xy - 6xz$.

From the first two terms factor an x^2, and from the last two terms factor $2x$. Then we have: $x^2(y - 3z) + 2x(y - 3z) = (y - 3z)(x^2 + 2x) = (y - 3z)(x + 2)x$.

20. Factors of $(a^n - b^n)$ and $(a^n + b^n)$.

Before leaving the general topic of factoring, we should make note of several conditions which determine the factors of the two forms $(a^n - b^n)$ and $(a^n + b^n)$.

We shall find that the factors in these cases depend sometimes on "n" being an odd or an even number.

(1) For all integral values of n,

$$a^n - b^n = (a - b)(a^{n-1} + a^{n-2}b + a^{n-3}b^2 + - - - - - + ab^{n-2} + b^{n-1}).$$

That is: $a^n - b^n$ is always divisible by $(a - b)$ and the second factor has all coefficients equal to $+1$. Thus,
$x^5 - y^5 = (x - y)(x^4 + x^3y + x^2y^2 + xy^3 + y^4)$.
Also, $a^4 - b^4 = (a - b)(a^3 + a^2b + ab^2 + b^3)$.

This last expression can be factored another way, since it can be written: $a^4 - b^4 = (a^2)^2 - (b^2)^2$

$$= (a^2 - b^2)(a^2 + b^2)$$
$$= (a - b)(a + b)(a^2 + b^2).$$

The reader should verify that in the first method shown above, $a^3 + a^2b + ab^2 + b^3 = (a + b)(a^2 + b^2)$, so that in the end the same result may be obtained.

(2) For an even value of n,

$$a^n - b^n = (a + b)(a^{n-1} - a^{n-2}b + a^{n-3}b^2 - a^{n-4}b^3 + \cdots - b^{n-1}).$$

That is: $a^n - b^n$ is always divisible by $(a + b)$ and the second factor has coefficients alternately $+1$ and -1. Thus $a^4 - b^4 = (a + b)(a^3 - a^2b + ab^2 - b^3)$, and the reader should verify that in this case the second factor can be decomposed so that:

$a^4 - b^4 = (a + b)(a - b)(a^2 + b^2)$, as before.

(3) For an odd value of n,

$$a^n + b^n = (a + b)(a^{n-1} - a^{n-2}b + a^{n-3}b^2 \cdots - ab^{n-2} + b^{n-1}).$$

Thus, $x^5 + y^5 = (x + y)(x^4 - x^3y + x^2y^2 - xy^3 + y^4)$.

(4) There is NO value of n for which $a^n + b^n$ will be divisible by $(a - b)$.

21. Highest Common Factor.

Certain numbers or expressions have no factors in common. Thus 3 and 7 have no common factor. Also, $x + a$ and $x - a$ have no factor in common. Such expressions are said to be **prime** to each other, or **relatively prime.**

Certain other expressions have one or more prime factors in common. Thus 6 and 15 have a common factor 3. Also $x^2 - a^2$ and $x^3 - a^3$ have a common factor $x - a$.

In many instances one is interested in finding the largest expression which is common to two or more expressions*. This greatest factor of several expressions is called the **highest common factor** and is usually abbreviated by H.C.F.

A simple way to find the H.C.F. is to write each expression as a product of its prime factors, then form the product of all the common prime factors. This product is the H.C.F.

* For algebraic expressions one finds the factor of highest degree.

Example 1. Find the H.C.F. of 36, 54, and 90.

Solution: $36 = 2 \cdot 2 \cdot 3 \cdot 3$

$54 = 2 \cdot 3 \cdot 3 \cdot 3$

$90 = 2 \cdot 3 \cdot 3 \cdot 5.$

In this case 2 is a common factor and 3 appears twice as a common factor. The product $2 \cdot 3 \cdot 3$ or 18 is the H.C.F.

Example 2. Find the H.C.F. of $x^2 - a^2$, $x^2 - 2ax + a^2$, and $x^3 - a^3$.

Solution: The prime factors of each expression are:

$$x^2 - a^2 = (x - a)(x + a)$$
$$x^2 - 2ax + a^2 = (x - a)(x - a)$$
$$x^3 - a^3 = (x - a)(x^2 + ax + a^2).$$

The only factor common to all three is $(x - a)$; hence the H.C.F. $= (x - a)$.

22. Lowest Common Multiple.

The **lowest common multiple** of two or more expressions is defined to be the least quantity which contains the given quantities as factors.

In order to find the L.C.M. one proceeds as follows. First, write each of the quantities as a product of its prime factors*. Then form a product which contains each distinct prime factor as many times as the maximum number of times it occurred in any one of the given quantities.

Example 1. Find the L.C.M. of 12, 15, 18, 21.

Solution: The numbers may be written as a product of the factors, thus:

$$12 = 2 \cdot 2 \cdot 3$$
$$15 = 3 \cdot 5$$
$$18 = 2 \cdot 3 \cdot 3$$
$$21 = 3 \cdot 7.$$

Since 2 occurs as a factor twice in 12,
Since 3 occurs as a factor twice in 18,
Since 5 occurs as a factor only once in 15,
Since 7 occurs as a factor only once in 21,
the L.C.M. $= 2 \cdot 2 \cdot 3 \cdot 3 \cdot 5 \cdot 7 = 1260.$

* For algebraic expressions one writes the irreducible linear, quadratic, etc , real factors.

Example 2. Find the L.C.M. of $a^2 - 2a - 15$, $a^2 + 2a - 35$, and $a^2 + 10a + 21$.

Solution: These may each be factored as,

$$a^2 - 2a - 15 = (a - 5)(a + 3)$$
$$a^2 + 2a - 35 = (a - 5)(a + 7)$$
$$a^2 + 10a + 21 = (a + 3)(a + 7).$$

Hence the L.C.M. $= (a - 5)(a + 3)(a + 7)$.

One usually leaves the L.C.M. in the product of factors form rather than perform the indicated multiplication.

Example 3. Find the L.C.M. of $x^2 - a^2$, $x^2 + 2ax + a^2$, and $x^3 - a^3$.

Solution: The factors of each are:

$$x^2 - a^2 = (x - a)(x + a).$$
$$x^2 + 2ax + a^2 = (x + a)(x + a).$$
$$x^3 - a^3 = (x - a)(x^2 + ax + a^2).$$

Hence the L.C.M. $= (x - a)(x + a)^2(x^2 + ax + a^2)$.

CHAPTER III

FRACTIONS

23. Introduction.

A **fraction** is an indicated division. Thus 3 divided by 4 is usually written $\frac{3}{4}$. In general, a/b is the notation for a fraction, where **b** is not equal to zero. The part "a" of the fraction is called the numerator, and "b" is called the denominator. Two fractions which have the same value are said to be **equivalent.** Thus $\frac{3}{4}$ and $\frac{9}{12}$ are equivalent. The fraction $\frac{9}{12}$ may be obtained from $\frac{3}{4}$ by multiplying both numerator and denominator by 3. Conversely, $\frac{9}{12}$ may be reduced to the form $\frac{3}{4}$ by dividing both numerator and denominator by 3. This latter we call **reduction to lowest terms,** and a fraction so reduced is said to be **simplified.**

Finally, it is to be understood that the three fractions $-\frac{2}{3}, \frac{-2}{3}, \frac{2}{-3}$ are all equivalent.

24. Addition and Subtraction of Fractions.

As in arithmetic, fractions may be added and subtracted, after they have been reduced to the same denomination. Thus,

$$\frac{1}{2} + \frac{3}{4} - \frac{2}{3} = \frac{6}{12} + \frac{9}{12} - \frac{8}{12} = \frac{6 + 9 - 8}{12} = \frac{7}{12}.$$

The reader will note that the process of reducing fractions to a common denomination involves the step of finding the L.C.M. of the separate denominators. For example, in the above illustration the L.C.M. of 2, 4, 3 is 12. The fraction $\frac{1}{2}$ may be changed to 12ths by multiplying both numerator and denominator by 6. Or one may change $\frac{1}{2}$ to $\frac{6}{12}$ by dividing the denominator 2 into 12 and obtaining 6, then multiplying the numerator 1 by 6.

Example 1. Perform the indicated operations:

$$\frac{1}{x - a} + \frac{1}{x + a} - \frac{2a}{x^2 - a^2}$$ and reduce to lowest terms.

Solution:

The L.C.M. of $(x - a)$, $(x + a)$ and $x^2 - a^2$ is $(x - a)(x + a)$ or $x^2 - a^2$.

This then must be used as the common denominator, and we have:

$$\frac{1}{x-a} + \frac{1}{x+a} - \frac{2a}{x^2-a^2} = \frac{x+a}{x^2-a^2} + \frac{x-a}{x^2-a^2} - \frac{2a}{x^2-a^2} =$$

$$\frac{x+a+x-a-2a}{x^2-a^2} = \frac{2x-2a}{x^2-a^2} = \frac{2(x-a)}{x^2-a^2} =$$

$$\frac{2\cancel{(x-a)}}{(x+a)\cancel{(x-a)}} = \frac{2}{x+a}.$$

This last step may be accomplished in either of two ways. Both numerator and denominator may be divided by $x - a$, or the actual division of the factor $(x - a)$ of the denominator into the $(x - a)$ of the numerator may be performed. This procedure is sometimes called **cancelling,** yet the reader should realize that cancelling is not an operation of algebra, but rather the result of actual division.

Example 2: Perform the indicated operations $\frac{5}{3-2x} + \frac{8}{2x-3} - \frac{3x-4}{2x^2-x-3}$ and reduce the result to lowest terms.

Solution:

The L.C.M. of the denominators is $(2x - 3)(x + 1)$. Note that $3 - 2x$ is contained in this common denominator $-(x + 1)$ times. Hence the numerator of the first fraction must be multiplied by $-(x + 1)$ when we write the fraction with the L.C.M. as a denominator.

Writing the above fractions with a common denominator gives:

$$\frac{5}{3-2x} + \frac{8}{2x-3} - \frac{3x-4}{(2x-3)(x+1)}$$

$$= \frac{-(x+1)5 + (x+1)8 - (3x-4)}{(2x-3)(x+1)}$$

$$= \frac{-5x - 5 + 8x + 8 - 3x + 4}{(2x-3)(x+1)}$$

$$= \frac{7}{(2x-3)(x+1)}.$$

25. Equivalent Fractions.

Rule 1.

If a fraction is obtained from a given fraction by multiplying both numerator and denominator by the same number, the two fractions are said to be **equivalent.** Thus $\frac{3}{4} = \frac{3}{4} \cdot \frac{5}{5} = \frac{15}{20}$ are equivalent, and $a/b = a/b \cdot c/c = ac/bc$ are equivalent.

Rule 2.

If a fraction is obtained from a given fraction by dividing both numerator and denominator by the same number "n," the two fractions are equivalent. (The number "n" must not be zero.)

26. Products of Fractions.

Rule 1.

The product of two or more fractions is a fraction whose numerator is the product of the separate numerators and whose denominator is the product of the separate denominators.

Example 1: Find the product of $\frac{2}{3}$ and $\frac{5}{7}$.
This may be written $\frac{2}{3} \cdot \frac{5}{7}$ or $\frac{2}{3} \times \frac{5}{7}$; $\frac{2}{3} \cdot \frac{5}{7} = \frac{10}{21}$.

Example 2: Find the product $\dfrac{9x^2 - 1}{x^2 - 25} \cdot \dfrac{x^2 + 5x}{9x - 3}$,

and reduce the result to the lowest terms.
One proceeds in such a problem by first writing each expression in factor form. Thus:

$$\frac{9x^2 - 1}{x^2 - 25} \cdot \frac{x^2 + 5x}{9x - 3} = \frac{\cancel{(3x - 1)}(3x + 1)}{(x - 5)\cancel{(x + 5)}} \cdot \frac{x\cancel{(x + 5)}}{3\cancel{(3x - 1)}} = \frac{x(3x + 1)}{3(x - 5)}.$$

This last step follows from the preceding form by dividing both numerator and denominator by $(3x - 1)(x + 5)$.

Note: A result involving fractions is said to be **reduced to lowest terms** (or simplified) whenever all common factors of the numerator and denominator have been removed by actual division.

27. Quotients of Fractions.

Before stating a rule for the quotient of two fractions, we define what is meant by the reciprocal of a number.

Definition: The **reciprocal** of any number "a" is unity divided by the number "a." Thus, the reciprocal of 5 is $\frac{1}{5}$, the reciprocal of $\frac{2}{3}$ is $\frac{3}{2}$. The formation of the reciprocal of a fraction merely inverts the fraction.

Rule.

To divide one fraction by another, multiply the dividend fraction by the reciprocal of the divisor fraction.

This rule results from the application of the rule for multiplication of fractions and of Equality Axiom Number 3. For, suppose that we let x represent the quotient obtained by dividing a/b by c/d. Thus:

$$\text{(A)} \qquad x = \frac{\frac{a}{b}}{\frac{c}{d}}.$$

Multiply both members of this equality by c/d.

$$\text{(B)} \qquad \frac{c}{d}x = \frac{\frac{a}{b}\left(\frac{c}{d}\right)}{\left(\frac{c}{d}\right)} = \frac{a}{b}.$$

Then multiply both members of this last equality by d/c.

$$\text{(C)} \qquad \frac{c}{d}\cdot\frac{d}{c}x = x = \frac{a}{b}\cdot\frac{d}{c}.$$

But from relations (A) and (C) we have:

$$x = \frac{\frac{a}{b}}{\frac{c}{d}} = \frac{a}{b}\cdot\frac{d}{c}.$$

Hence the rule.

Example 1: Perform the indicated division $\frac{16a^2b^3}{5ac} \div \frac{8ab^2}{15c^2}$ and simplify the result.

Solution:

$$\frac{\dfrac{16a^2b^3}{5ac}}{\dfrac{8ab^2}{15c^2}} = \frac{\overset{2}{\cancel{16}}a^2b^3}{\cancel{5}ac} \cdot \frac{\overset{3}{\cancel{15}}c^2}{\cancel{8}ab^2}$$

$$= \frac{6a^2b^3c^2}{a^2b^2c} = 6bc.$$

Example 2: Perform the division

$$\frac{x^2 - 5x + 6}{x + 5} \div \frac{x - 3}{x^2 + 8x + 15}$$

and simplify the result.

Solution: $\dfrac{x^2 - 5x + 6}{x + 5} \div \dfrac{x - 3}{x^2 + 8x + 15} =$

$$\frac{x^2 - 5x + 6}{x + 5} \cdot \frac{x^2 + 8x + 15}{x - 3} = \frac{\cancel{(x - 3)}(x - 2)}{\cancel{x + 5}} \cdot \frac{\cancel{(x + 5)}(x + 3)}{\cancel{x - 3}}$$

$$= (x - 2)(x + 3)$$

$$= x^2 + x - 6.$$

This final result follows from the preceding one by making use of product type (E), Chapter II.

28. Complex Fractions.

Definition: A **complex fraction** is one whose numerator or denominator, or both, consists of fractions.

Such fractions are simplified whenever one has performed all of the indicated operations. The steps in the simplification are performed in the following order. First, perform the additions or subtractions; secondly, perform the indicated multiplications and divisions.

Example 1: Simplify the fraction: $\dfrac{\dfrac{x}{x - 1} - 1}{1 + \dfrac{x}{1 - x}}.$

Solution: The first step consists of performing the subtraction indicated in the numerator and the addition indicated in the denominator.

Thus: $$\frac{\frac{x}{x-1}-1}{1+\frac{x}{1-x}} = \frac{\frac{x-(x-1)}{x-1}}{\frac{(1-x)+x}{1-x}} = \frac{\frac{x-x+1}{x-1}}{\frac{1-x+x}{1-x}}$$

$$= \frac{\frac{1}{x-1}}{\frac{1}{1-x}} = \frac{1}{x-1} \cdot \frac{1-x}{1} = -1.$$

(Note that $x - 1$ is contained in $1 - x$ minus one times, as can be seen by multiplying $x - 1$ by -1.)

Example 2: Simplify: $\frac{x^2-y^2}{xy} \div \left(\frac{1}{x}+\frac{1}{y}\right)$.

Solution:

$$\frac{\frac{x^2-y^2}{xy}}{\frac{1}{x}+\frac{1}{y}} = \frac{\frac{x^2-y^2}{xy}}{\frac{x+y}{xy}} = \frac{x^2-y^2}{xy} \cdot \frac{xy}{x+y} = x - y.$$

The last step follows from product form (C), Chapter II.

Example 3: Simplify: $x - \frac{2x}{x^2 - \frac{x}{1+\frac{2}{1-x}}}$.

Solution: Such a problem as this involves several separate steps. One begins the simplification by starting with the last indicated division and proceeding to each successive division, at the same time simplifying each step in turn. Thus we begin by adding $1 + \frac{2}{1-x}$.

$$1 + \frac{2}{1-x} = \frac{1-x+2}{1-x} = \frac{3-x}{1-x}.$$

The next indicated step is to divide x by $\frac{3-x}{1-x}$.

$$\frac{x}{\frac{3-x}{1-x}} = x \cdot \frac{(1-x)}{3-x} = \frac{x-x^2}{3-x}.$$

This result must be subtracted from x^2.

$$x^2 - \frac{x - x^2}{3 - x} = \frac{x^2(3 - x) - (x - x^2)}{3 - x} = \frac{3x^2 - x^3 - x + x^2}{3 - x}$$

$$= \frac{4x^2 - x^3 - x}{3 - x} = \frac{x(4x - x^2 - 1)}{3 - x}.$$

Now 2x must be divided by this last result.

$$\frac{2x}{\frac{x(4x - x^2 - 1)}{3 - x}} = 2x \cdot \frac{(3 - x)}{x(4x - x^2 - 1)} = \frac{2(3 - x)}{4x - x^2 - 1}.$$

Finally, this latest result must be subtracted from x.

$$x - \frac{2(3 - x)}{4x - x^2 - 1} = \frac{x(4x - x^2 - 1) - 2(3 - x)}{4x - x^2 - 1}$$

$$= \frac{4x^2 - x^3 - x - 6 + 2x}{4x - x^2 - 1}$$

$$= \frac{-x^3 + 4x^2 + x - 6}{-x^2 + 4x - 1} = \frac{x^3 - 4x^2 - x + 6}{x^2 - 4x + 1}.$$

CHAPTER IV

EXPONENTS

29. Definitions.

We begin a discussion of exponents by assuming that the reader is familiar with the notation proposed in Chapter I, Sec. 10.

The symbol a^3 means $a \cdot a \cdot a$. The index 3 is the exponent, which indicates the number of a's in the product. On this basis we are in a position to give the following definition.

Definition: The symbol a^m, where **m** is a **positive integer,** represents the product of **m** factors each of which is **a**. The integer **m** is the index, or **exponent,** which represents the number of times the **base** number "a" appears as a factor.

Illustration:

$$a^5 = a \cdot a \cdot a \cdot a \cdot a; \quad 2^3 = 2 \cdot 2 \cdot 2 = 8.$$

30. Laws of Exponents.

It is necessary for us to understand the meaning of such expressions as $2^3 \cdot 2^2$, and to be able to evaluate such expressions. This understanding follows at once from the definition of an exponent and the definition of multiplication. Thus $2^3 \cdot 2^2 = (2 \cdot 2 \cdot 2) \cdot (2 \cdot 2) = 2^5$ by definition.

The idea involved in this illustration can be generalized, and gives rise to a law of operation for use in the theory of exponents.

Law 1. $\mathbf{a^m \cdot a^n = a^{m+n}}$.

Proof: $a^m \cdot a^n = \underbrace{(a \cdot a \cdot a \cdots a)}\underbrace{(a \cdot a \cdot a \cdots a)}$

$= \mathbf{m}$ factors **a,** followed by **n** factors **a,**

or a total of $m + n$ factors **a**.

Hence by the definition of exponent $\mathbf{a^m \cdot a^n = a^{m+n}}$.

This law may be stated in words as follows: "The product of two different powers of the same number (or base) is equal to the base number raised to a power which is the sum of the exponents."

By extending this law to a product of three or more components we arrive at such expressions as:

$$a^m \cdot a^n \cdot a^r = a^{m+n+r}, \text{etc.}$$

Examples: $2^3 \cdot 2^4 = 2^{3+4} = 2^7 = 128.$

$x^3 \cdot x^3 = x^{3+3} = x^6.$

Law 2. $(a^m)^n = a^{mn}$.

Proof: By definition this means

$$(a^m)^n = \underbrace{a^m \cdot a^m \cdot a^m \cdot \cdot \cdot \cdot a^m}_{\text{n of these factors}}$$

$$= (a \cdot a \cdot a \cdot \cdot \cdot \cdot a) \cdot (a \cdot a \cdot a \cdot \cdot \cdot \cdot a) \cdot \cdot \cdot \cdot (a \cdot a \cdot a \cdot \cdot \cdot \cdot a)$$

$= a^{mn}$, since there are **n** such parentheses each containing **m** factors a; the total being **mn** factors **a.**

Examples: $(2^2)^3 = 2^{2 \cdot 3} = 2^6 = 64.$

$(x^3)^3 = x^{3 \cdot 3} = x^9.$

Law 3. (a) $(a \cdot b)^n = a^n b^n$.

Proof: $(a \cdot b)^n = \underbrace{(ab)(ab)(ab) \ . \ . \ . \ (ab)}_{\text{n such factors}}$

(by definition).

On rearranging the letters of this product by writing first the **n** factors "a," followed by the **n** factors "b," we have:

$$(ab)^n = (a \cdot a \cdot a \cdot \cdot \cdot \cdot \cdot \cdot a)(b \cdot b \cdot b \cdot \cdot \cdot \cdot \cdot \cdot b)$$
$$= a^n b^n.$$

Example: $(2x)^3 = 2^3 \cdot x^3 = 8x^3.$

(b) $(a \cdot b \cdot c \cdot \cdot \cdot \cdot \cdot)^n = a^n \cdot b^n \cdot c^n \cdot \cdot \cdot \cdot \cdot$

This means that Law 3 can be extended to cover any finite number of factors.

Law 4. (a) $\dfrac{a^m}{a^n} = a^{m-n}$. If m is greater than n.

(b) $\dfrac{a^m}{a^n} = \dfrac{1}{a^{n-m}}$. If n is greater than m.

Proof of (a). Consider first an illustration where m and n are definite integers, such as:

$\frac{a^5}{a^2} = \frac{\not a \cdot \not a \cdot a \cdot a \cdot a}{\not a \cdot \not a} = a^{5-2} = a^3$ since the a's in the denominator will cancel an equal number of a's in the numerator and leave a residual number of factors namely $5 - 2 = 3$.

$$\text{In general } \frac{a^m}{a^n} = \frac{(a \cdot a \cdot a \cdots \text{ to m factors})}{(a \cdot a \cdot a \cdots \text{ to n factors})} =$$

a^{m-n}, and as noted above **n** of these factors cancel leaving $m - n$ factors "a" in the numerator.

Case (b). The proof of this case is similar to that for case (a) with the exception that the m factors of the numerator cancel a like number of the denominator leaving $n - m$ factors "a" in the denominator.

$$\text{Examples: } \frac{2^5}{2^3} = 2^{5-3} = 2^2 = 4.$$

$$\frac{a^3}{a^6} = \frac{1}{a^{6-3}} = \frac{1}{a^3}.$$

$$\frac{2^3}{2^5} = \frac{1}{2^{5-3}} = \frac{1}{2^2} = \frac{1}{4}.$$

Law 5. $\left(\frac{a}{b}\right)^n = \frac{a^n}{b^n}.$

Proof: Again by definition of an exponent,

$$\left(\frac{a}{b}\right)^n = \left(\frac{a}{b}\right)\left(\frac{a}{b}\right)\left(\frac{a}{b}\right) \cdots\cdots \left(\frac{a}{b}\right), \left[\text{n factors } \frac{a}{b}\right].$$

By the rule for the product of fractions, Sec. 26, we have: $\left(\frac{a}{b}\right)^n = \frac{(a \cdot a \cdot a \cdots a)}{(b \cdot b \cdot b \cdots b)} = \frac{a^n}{b^n}.$

$$\text{Examples: } \left(\frac{2}{3}\right)^2 = \frac{2^2}{3^2} = \frac{4}{9}.$$

$$\left(\frac{2a}{b}\right)^2 \cdot \left(\frac{b}{6a}\right)^2 \cdot \left(\frac{3a^2}{2b}\right)^3 = \frac{2^2\not a^2}{\not b^2} \cdot \frac{\not b^2}{6^2\not a^2} \cdot \frac{3^3a^6}{2^3b^3} =$$

$$= \frac{4 \cdot 27a^6}{36 \cdot 8b^3} = \frac{3a^6}{8b^3}.$$

31. Negative Exponents.

In the discussion of Law 4 for positive exponents we had to make a distinction of two cases depending on whether **m** was greater or less than **n**. This restriction can be removed if we allow ourselves the use of negative exponents, and assume that the first part of law 4 holds whether **m** is greater than or less than **n**. With this removal of restrictions Law 4(a) gives $\frac{a^3}{a^6} = a^{3-6} = a^{-3}$, but this has been shown in the second example under Law 4 to be also equal to $\frac{1}{a^3}$. Hence $\frac{1}{a^3} = a^{-3}$, and we may extend our notion of exponents in keeping with this result. We now have

Law 6. $\mathbf{a^{-n} = \frac{1}{a^n}}$ **or** $\mathbf{\frac{1}{a^{-n}} = a^n}$.

Another way of expressing Law 6 is to say that any factor may be changed from the numerator into the denominator, or vice versa, provided one changes the sign of the exponent.

32. Zero Exponent.

Using the base "a" as any quantity except zero, we now examine the case under Law 4 where $m = n$.

If Law 4 is to hold, then we must have $\frac{a^n}{a^n} = a^{n-n} = a^0$.

But since any quantity divided by itself is unity, we have $\frac{a^n}{a^n} = 1$.

Consequently, it follows that $\frac{a^n}{a^n} = a^0 = 1$. (By Axiom 5, Sec. 14.)

Consequently, we define

Law 7. $\mathbf{a^0 = 1}$**, if** $\mathbf{a \neq 0}$**.**

Illustration 1.

$3^0 = 1.$

Illustration 2.

$(3x)^0 = 1.$

Illustration 3.

$3x^0 = 3.$

The reader's attention is called to the fact that the last two illustrations are quite different. In the case $(3x)^0$, the entire product 3x is raised to the zero power, thus having the value 1 for all values of x which are different from zero. (For if $x = 0$, then the base 3x becomes 0, and this case was barred throughout our discussion. The case where the base is zero leads to the question of indeterminate forms, a topic which requires some knowledge of the theory of limits, and which is usually discussed in considerable detail in a course in calculus.)

For the last illustration $3x^0$, we must note that the x only is raised to the zero power. Consequently, we have $3x^0 = 3 \cdot 1 = 3$. The student should exert care to distinguish between these two cases in solving problems. They are but special cases of the following:

Illustration 4.

$$(2x)^3 = 2^3 \cdot x^3 = 8x^3.$$

Illustration 5.

$$2x^3 = 2 \cdot x^3.$$

33. Fractional Exponents.

The seven laws of exponents also apply in case the exponents are fractions. In this present chapter, we shall illustrate only how these laws apply to fractional exponents. For a more complete discussion of the relation between fractional exponents and extraction of roots, see Chapter V.

Illustration 1.

By Law 1.

$$a^{2/3} \cdot a^{3/4} = a^{2/3 + 3/4} = a^{8/12 + 9/12} = a^{17/12}.$$

Illustration 2.

By Law 2.

$$(a^{2/3})^{3/4} = a^{2/3 \cdot 3/4} = a^{6/12} = a^{1/2}.$$

Illustration 3.

$$\frac{a^{2/3}}{a^{3/4}} = a^{2/3 - 3/4} = a^{8/12 - 9/12} = a^{-1/12} = \frac{1}{a^{1/12}}.$$

(By Law 4 followed by Law 6.)

The student should note that in applying the laws of exponents to cases in which the exponents are fractions, care must be taken to apply the rules for combining fractions as given in Chapter III.

Chapter V

RADICALS

34. Introduction.

An expression such as $\sqrt[n]{a}$ is called a **radical.** The number **n** is called the **index,** and the number **a** is called the **radicand.**

If the index $n = 2$, then $\sqrt[2]{a}$ is called the square root of **a.** Usually the index 2 is omitted, and the expression is simply written $\sqrt{a}$.

> **Definition 1:** The **square root** of a number **a** is that value which multiplied by itself gives a product equal to **a.**
>
> **Definition 2:** The **cube root** of a number **a** is that value which used three times as a factor gives a product equal to **a.**

In general $\sqrt[n]{a}$ represents the nth root of a number **a.**

> **Definition 3:** Two radicals are said to be the **same** if they have the same index and radicand.
>
> **Definition 4:** Two radicals are said to be **similar** if, upon reduction to their simplest form, their radicals are the same.
>
> Thus $\sqrt{2}$ and $\frac{1}{3}\sqrt{2}$ are similar radicals, but $\sqrt{2}$ and $\sqrt[3]{2}$ are not similar, since the indices are different. Also $\sqrt{2}$ and $\sqrt{3}$ are **not** similar, since the radicands are different.

35. Principal Roots.

By the nth root of a number "a" we mean a number whose n^{th} power is equal to **a.** Thus $2^4 = 16$, and $(-2)^4 = 16$, so that it is rather obvious that a number can have more than one n^{th} root. (In the chapter on the Theory of Equations we shall see that a number has **n**, n^{th} roots.) For our purposes we must restrict our notation in order that we may pick a particular root of **a** given number.

If **n** is an odd integer and **a** is a real number, then there is only one real n^{th} root. Thus, $\sqrt[3]{8} = 2$, $\sqrt[5]{-32} = -2$. We define $\sqrt[n]{a}$ under these circumstances to be this real n^{th} root and call it the **principal n^{th} root.**

If **n** is an even integer and **a** is a positive number, then **a** has two real n^{th} roots. Thus, $\sqrt[4]{16} = +2$, $-\sqrt[4]{16} = -2$. We define the positive one of these roots to be the **principal** value. In order to indicate the negative root we must affix the $-$ sign to the radical.

If **n** is an even integer and **a** is a negative number there are no real roots. See Chapter XVII.

> **Agreement.** When we write $\sqrt[4]{16}$, as above, we shall mean $+2$. If we wish to indicate the negative root, we shall write $-\sqrt[4]{16}$. On occasion, we may wish to indicate both values, in which case we shall write $\pm\sqrt[4]{16}$, or ± 2, to indicate the dual choice.

36. Relations between Radicals and Exponents.

Recalling the theory of exponents, we have:

$$a^{1/2} \cdot a^{1/2} = a^{1/2 + 1/2} = a.$$

But in the present chapter we have defined: $\sqrt{a} \cdot \sqrt{a} = a$.

If both these statements are to be true, one must identify $a^{1/2}$ with $\sqrt{a}$, by agreeing that $a^{1/2} = \sqrt{a}$. By a similar argument $a^{1/3} \cdot a^{1/3} \cdot a^{1/3} = a$. Also $\sqrt[3]{a} \cdot \sqrt[3]{a} \cdot \sqrt[3]{a} = a$. Hence we agree that $a^{1/3} = \sqrt[3]{a}$. Thus we see that a generalization would lead to the statement,

$$\mathbf{a^{1/n} = \sqrt[n]{a}.}$$

This equation allows one to establish the laws of radicals from the laws of exponents.

37. The Laws of Radicals.

By definition $(\sqrt[n]{a})^n = a$. This could also be arrived at from the laws of exponents. For, $\sqrt[n]{a}$ can be written $a^{1/n}$. Then by Law 2 of exponents, $(a^{1/n})^n = a^{n/n} = a$, giving therefore

Law 1.

$$\mathbf{(\sqrt[n]{a})^n = a.}$$

Illustrations.

$$(\sqrt[4]{5})^4 = 5.$$
$$(\sqrt[3]{-8})^3 = -8.$$

In order to find an expression for $\sqrt[n]{a} \cdot \sqrt[n]{b}$, we first write an equivalent expression in exponent form, $a^{1/n} \cdot b^{1/n}$. By Law 3 of exponents, $a^{1/n} \cdot b^{1/n} = (ab)^{1/n}$. Rewriting this last result as $\sqrt[n]{ab}$ of the radical notation, we have,

Law 2. $\quad \sqrt[n]{\mathbf{a}} \cdot \sqrt[n]{\mathbf{b}} = \sqrt[n]{\mathbf{ab}}.$

Illustration.

$$\sqrt[3]{2} \cdot \sqrt[3]{5} = \sqrt[3]{10}.$$

To find an expression for $\sqrt[m]{\sqrt[n]{a}}$, we write $\sqrt[m]{a^{1/n}} = (a^{1/n})^{1/m} = a^{1/mn} = \sqrt[mn]{a}$. Hence by the use of Law 2 of exponents we have,

Law 3. $\quad \sqrt[m]{\sqrt[n]{\mathbf{a}}} = \sqrt[mn]{\mathbf{a}}.$

Illustration.

$$\sqrt[3]{\sqrt[2]{5}} = \sqrt[6]{5}.$$

To find an expression for $\dfrac{\sqrt[n]{a}}{\sqrt[n]{b}}$ we write this in terms of exponents and apply Law 5 of exponents. Thus,

$$\frac{\sqrt[n]{a}}{\sqrt[n]{b}} = \frac{a^{1/n}}{b^{1/n}} = \left(\frac{a}{b}\right)^{1/n} = \sqrt[n]{\frac{a}{b}}.$$

Consequently, we have,

Law 4. $\quad \dfrac{\sqrt[n]{\mathbf{a}}}{\sqrt[n]{\mathbf{b}}} = \sqrt[n]{\dfrac{\mathbf{a}}{\mathbf{b}}}.$

Illustration.

$$\frac{\sqrt[3]{10}}{\sqrt[3]{5}} = \sqrt[3]{\frac{10}{5}} = \sqrt[3]{2}.$$

Finally, we find an expression for $\sqrt[q]{a^p}$.

$$\sqrt[q]{a^p} = (a^p)^{1/q} = a^{p/q},$$

and we have,

Law 5. $\quad \sqrt[q]{\mathbf{a}^{\mathbf{p}}} = \mathbf{a}^{\mathbf{p/q}}.$

This Law 5 tells us that the numerator **p** of a fractional exponent indicates the power to which **a** is to be raised, and that the denominator **q** indicates the root to be extracted.

Illustration.

$$\sqrt[3]{8^2} = \sqrt[3]{64} = 4 \text{ or}$$
$$\sqrt[3]{8^2} = (\sqrt[3]{8})^2 = 2^2 = 4.$$

Usually, one extracts the root before raising to the power, thus avoiding very large numbers in some instances.

38. Changes in the Radicand.

(*a*) Removing factors from the radicand.

Any factor which is a perfect n^{th} power can be removed from the radicand as shown in the following illustrations.

$$\sqrt{63} = \sqrt{9 \cdot 7} = \sqrt{3^2 \cdot 7} = 3\sqrt{7}.$$
$$\sqrt[3]{54x^4y^3} = \sqrt[3]{27x^3y^3 \cdot 2x} = 3xy\sqrt[3]{2x}.$$

(*b*) Introducing quantities under the radical.

Any coefficient may be introduced under the radical sign provided it is raised to a proper power corresponding to the root, before being made a part of the radicand. Thus

$$5\sqrt{3} = \sqrt{5^2} \cdot \sqrt{3} = \sqrt{25 \cdot 3} = \sqrt{75}.$$

$$2ab^2\sqrt[3]{7ax} = \sqrt[3]{2^3a^3b^6}\,\sqrt[3]{7ax} = \sqrt[3]{8a^3b^6 \cdot 7ax} = \sqrt[3]{56a^4b^6x}.$$

(*c*) Making the radicand integral.

If the radicand is a fraction, it may be made integral by methods illustrated in the following examples.

Illustrations.

$$\sqrt{\frac{2}{3}} = \sqrt{\frac{2}{3} \cdot \frac{3}{3}} = \sqrt{\frac{6}{3^2}} = \frac{\sqrt{6}}{3} = \frac{1}{3}\sqrt{6}.$$

$$\sqrt{\frac{r}{s}} = \sqrt{\frac{r}{s} \cdot \frac{s}{s}} = \sqrt{\frac{rs}{s^2}} = \frac{1}{s}\sqrt{rs}.$$

Such procedure is called **rationalizing the denominator** and will be treated in more detail later in this chapter. See Sec. 43.

(*d*) Reducing the index of the radical.

Many radicals may have the index reduced as shown by the following examples.

Illustrations.

$$\sqrt[4]{9x^2} = \sqrt[4]{(3x)^2} = (3x)^{2/4} = (3x)^{1/2} = \sqrt{3x}.$$

$$\sqrt[6]{-27} = \sqrt[6]{(-3)^3} = (-3)^{3/6} = (-3)^{1/2} = \sqrt{-3}.$$

The changes involved in cases a, c, d, above are said to simplify radicals. Thus we may set up a set of rules known as:

Criterions for Simplifying Radicals.

1. Remove all perfect n^{th} powers from the radicand. (See a.)
2. Make the radicand integral. (See c.)
3. Make the index of the radical as small as possible. (See d.)

Any radical so treated will then be in simplest form.

39. Rational and Irrational Numbers.

The process of extracting roots of numbers leads us to a type of number called **irrational.** But before we can define such numbers we must note the following definitions.

Definition 1: A **prime** number is one which has only itself and unity as divisors.

Illustration.

3, 5, 7, 17, etc., are prime numbers.

Definition 2: Two numbers are said to be **relatively prime** if they have no factors in common. (The factor unity is disregarded since unity is a factor of every number.)

Illustrations.

3 and 11 are relatively prime.

5 and 9 are relatively prime although 9 itself is not a prime number.

Definition 3: A **rational** number is one which can be expressed as the quotient of two integers which are relatively prime.

Illustrations.

6 is a rational number, for $6 = 6/1$.

$\frac{3}{4}$ is a rational number, for it is such a quotient.

$2\frac{2}{3}$ is a rational number, for $2\frac{2}{3} = \frac{8}{3}$.

.036 is a rational number, for $.036 = \frac{36}{1000} = \frac{9}{250}$.

Definition 4: An **irrational** number is a real number which is not rational.

Illustrations.

$\sqrt{2}$, $\sqrt[3]{2}$, $\sqrt{3}$, $\sqrt{5}$, $1 + \sqrt{6}$, etc. are all irrational numbers, for none of them can be expressed as the quotient of two relatively prime integers. It is not always an easy task to prove that a given number is irrational. However $\sqrt{2}$ is easily proved to be an irrational number. A proof is given in the appendix, see p. 205.

In the preceding section we talked about the process of rationalizing the denominator of a fractional radical form. We see now that such a procedure simply means that we make the denominator a rational number. The process also leads to a distinct advantage when evaluating the numerical value in terms of decimals, as is shown in the following example.

Illustration.

Evaluate $\frac{1}{\sqrt{2}}$ to three decimal places.

This can be done directly by dividing unity by 1.414, but the process involves long division. On the other hand we may write $\frac{1}{\sqrt{2}} = \frac{1}{\sqrt{2}} \cdot \frac{\sqrt{2}}{\sqrt{2}} = \frac{\sqrt{2}}{\sqrt{4}} = \frac{\sqrt{2}}{2}$. This is approximately $\frac{1.414}{2} = .707$ and involves only simple division. The reader may verify that $\frac{1}{1.414}$ also gives .707.

Consequently, one always rationalizes the denominator of fractional forms involving radicals before evaluating the form as a decimal.

40. Addition and Subtraction of Radicals.

Similar radicals may be added or subtracted the same as any other algebraic quantities.

Illustration 1.

$6\sqrt{2} + 5\sqrt{2} - 4\sqrt{2} = (6 + 5 - 4)\sqrt{2} = 7\sqrt{2}.$

(If the sequence of terms all involve the same radicand,

the sums and differences of the coefficients are obtained and the result is then multiplied by the common radical factor.)

If the radicals involved do not have the same radicand, the sum or difference may be obtained provided the simplification of the radicals reduces all of them to forms involving a common radicand. Thus,

Illustration 2.

$$3\sqrt{a^3b^3} + \sqrt{4ab^3} - 6a\sqrt{a^3b} =$$
$$3ab\sqrt{ab} + 2b\sqrt{ab} - 6a^2\sqrt{ab} =$$
$$(3ab + 2b - 6a^2)\sqrt{ab}.$$

Dissimilar radicals cannot be added. The most that one can do is to indicate the sum. Thus, $\sqrt{3} - \sqrt[3]{5} + \sqrt[3]{3}$ cannot be further combined.

41. Radicals of Different Indices.

Radicals of different indices may be reduced to radicals of the same index. The common index for the several radicals is chosen to be the L.C.M. of the given indices.

Illustration 1.

Write $\sqrt{2}$, $\sqrt[3]{4}$, $\sqrt[4]{5}$ as radicals of the same index. The indices are 2, 3, and 4; their L.C.M. is 12, so that each may be expressed as a radical of index 12:

$$\sqrt{2} = 2^{1/2} = 2^{6/12} = \sqrt[12]{2^6} = \sqrt[12]{64}.$$
$$\sqrt[3]{4} = 4^{1/3} = 4^{4/12} = \sqrt[12]{4^4} = \sqrt[12]{256}.$$
$$\sqrt[4]{5} = 5^{1/4} = 5^{3/12} = \sqrt[12]{5^3} = \sqrt[12]{125}.$$

It should be pointed out that $\sqrt{2}$ and $\sqrt[12]{64}$ are not equivalent in every respect. The $\sqrt{2}$ has two roots, the $\sqrt[12]{64}$ has 12 roots, as we shall see in Chapter XVIII, on the Theory of Equations. Among the 12 values in this second case there are the two square roots of 2. The **principal** value in either case would be the same.

It should also be noted that in the first illustration of part (d), Sec. 38, we had $\sqrt[4]{9x^2} = \sqrt{3x}$. The first part involves 4 fourth roots, the second 2 square roots which are among the four roots of the left hand member. However, since the expressions involve a variable quantity x, one notes that if $x = -1$, the left member gives $\sqrt[4]{9}$, and has a real root, but $\sqrt{-3}$ has no real roots. (See Chapter XVII.)

Therefore, in changing the index of a radical, care must be exercised to choose particular roots for which both expressions are true and equal. In the above example the values of the variable x must be limited therefore to positive values, if some pair of roots are to be equal.

Reduction of radicals to the same index also helps answer such questions as:

$$\text{Is } \sqrt{32} \text{ greater or less than } \sqrt[3]{181}?$$

If one should extract the roots in each case to three decimal places, obtaining in each case 5.656, it might appear that these values are the same. However, if one writes $\sqrt{32}$ as $\sqrt[6]{32^3} = \sqrt[6]{32768}$ and $\sqrt[3]{181}$ as $\sqrt[6]{181^2} = \sqrt[6]{32761}$, then the question is easily settled; for one recognizes at a glance that since 32768 is greater than 32761, then $\sqrt{32}$ is greater than $\sqrt[3]{181}$. As a matter of fact, these roots differ only slightly in the fourth decimal place, being respectively 5.6567 and 5.6566.

42. Products of Radicals.

To find the product of two or more radicals of the same index, multiply the coefficients to obtain the coefficient of the product, and multiply the radicals by means of $\sqrt[n]{a} \cdot \sqrt[n]{b} = \sqrt[n]{ab}$ to obtain the radical of the product.

Illustrations.

(1) $2\sqrt{5} \cdot 3\sqrt{2} = 6\sqrt{10}$.

(2) $5\sqrt{2} \cdot 6\sqrt{8} = 30\sqrt{16} = 120$. (Note that the product of two or more radicals may be a rational quantity, as in this example.)

To find the product of two or more radicals of different indices, first reduce to radicals of the same index and proceed as above.

Illustration.

$$3\sqrt[3]{2} \cdot 5\sqrt{3} = 3\sqrt[6]{4} \cdot 5\sqrt[6]{27} = 15\sqrt[6]{108}.$$

43. Division of Radicals.

This section will be divided into two parts. The procedure will be shown by examples.

Part A. Simple Expressions.

Illustration 1.

Perform the division $\frac{6}{\sqrt{3}}$.

This we write as $\frac{6}{\sqrt{3}} \cdot \frac{\sqrt{3}}{\sqrt{3}} = \frac{6\sqrt{3}}{3} = 2\sqrt{3}$.

Illustration 2.

Simplify $\sqrt{\frac{5}{6}} \div \sqrt{\frac{15}{8}}$.

$$\sqrt{\frac{5}{6}} \div \sqrt{\frac{15}{8}} = \sqrt{\frac{\frac{5}{6}}{\frac{15}{8}}} = \sqrt{\frac{5}{6} \cdot \frac{8}{15}} = \sqrt{\frac{4}{9}} = \frac{2}{3}.$$

Rule: The indicated division $\frac{a}{\sqrt{k}}$ can always be performed by multiplying both numerator and denominator by $\sqrt{k}$, leaving the result with a rational denominator.

Definition: When two binomial expressions involving radicals differ only in the sign between the terms, they are called **conjugate**. Thus, $\sqrt{a} + \sqrt{b}$ and $\sqrt{a} - \sqrt{b}$ are each the conjugate of the other. Also, $2 + \sqrt{3}$ and $2 - \sqrt{3}$ are each the conjugate of the other.

Part B. Compound Expressions.

Illustration 3.

Perform the division $\frac{7}{5 - 3\sqrt{2}}$.

This division is performed by rationalizing the denominator. To do this, multiply both numerator and denominator by the conjugate of $5 - 3\sqrt{2}$, and simplify the result.

Thus,

$$\frac{7}{5 - 3\sqrt{2}} \cdot \frac{5 + 3\sqrt{2}}{5 + 3\sqrt{2}} = \frac{7(5 + 3\sqrt{2})}{25 - 18} = \frac{7(5 + 3\sqrt{2})}{7}$$
$$= 5 + 3\sqrt{2}.$$

Illustration 4.

Write $\dfrac{3\sqrt{5}-\sqrt{3}}{\sqrt{5}+2\sqrt{3}}$ in an equivalent form with rational denominator.

$$\frac{3\sqrt{5}-\sqrt{3}}{\sqrt{5}+2\sqrt{3}} = \frac{3\sqrt{5}-\sqrt{3}}{\sqrt{5}+2\sqrt{3}} \cdot \frac{\sqrt{5}-2\sqrt{3}}{\sqrt{5}-2\sqrt{3}}$$

$$= \frac{3\cdot 5 - 6\sqrt{15} - \sqrt{15} + 2\cdot 3}{5-12}$$

$$= \frac{21 - 7\sqrt{15}}{-7} = \sqrt{15} - 3.$$

Rule: In general an expression of the form $\dfrac{r+s}{\sqrt{a}\pm\sqrt{b}}$ may be rationalized by multiplying both numerator and denominator by the conjugate of the expression in the denominator of the given quotient.

If three or more radicals appear in the denominator, one may proceed to rationalize by first grouping the terms and applying the general principle two or more times. Thus $\dfrac{1}{\sqrt{a}+\sqrt{b}-\sqrt{c}}$ may be written $\dfrac{1}{(\sqrt{a}+\sqrt{b})-\sqrt{c}}$ and the rationalizing factor $(\sqrt{a}+\sqrt{b})+\sqrt{c}$ applied. This will reduce the form to one containing but two radicals, and then previous rules will apply.

44. Equations Involving Radicals.

A discussion of such equations will be found in Chapter XII, Sec. 88.

CHAPTER VI

RATIO, PROPORTION, AND VARIATION

45. Ratio and Proportion.

In elementary algebra and geometry we studied the theory of ratio and proportion. We list now some of the basic ideas in order to prepare the way for the study of variation.

Definition 1: The **ratio** of two quantities **a** to **b** is the quotient $a \div b$, or $a : b$, or a/b. (We shall use the last notation.)

Definition 2: A **proportion** is the equality of two ratios.

Thus $a/b = c/d$ is a proportion.

In this proportion:

1) b and c are called the **means.**
2) a and d are called the **extremes.**
3) d is called the **fourth proportional** to a, b, and c.
4) If $b = c$ then $a/c = c/d$ and c is called the **mean proportional** between a and d.

Two important facts should be noted. If $a/b = 3/4$, this does not mean that $a = 3$ and $b = 4$. In fact, we might have $a = 9$, $b = 12$, or any other set of values which may be reduced to the fraction $3/4$.

The ratio of two quantities does not change if both quantities are measured in the same units. For example, the ratio of 3 feet to 1 foot is the same as the ratio of 36 inches to 12 inches.

46. Properties of Proportions.

If $a/b = c/d$, then

1) $ad = bc$. (The product of the means = product of the extremes.)
2) $a/c = b/d$. (By alternation.)
3) $\frac{a + b}{b} = \frac{c + d}{d}$.

This is obtained as follows: add 1 to each side of the given proportion: $a/b + 1 = c/d + 1$. Reducing to a common denominator gives (3).

4) $\frac{a - b}{b} = \frac{c - d}{d}$. (Hint: subtract 1 from each side, etc.)

5) $\frac{a + b}{a - b} = \frac{c + d}{c - d}$. (By dividing 3 by 4.)

47. Variation.

Let us consider a special case of proportion where $y/x = k/1$, or more simply $y/x = k$.

Now suppose that x and y are allowed to vary and that **k** is a constant. If y has the value 2 and x the value 1, then $k = 2$. If now we wish to keep $k = 2$, and let x and y vary, x must increase when y increases and decrease when y decreases. That is, they must vary in the same way. Such a type of variation is called direct.

Definition 1: If x and y are variables and **k** is a constant, then $y/x = k$ or $y = kx$ represents **direct variation. k** is called the factor of proportionality.

Definition 2: If y varies directly as x^n, then $y = kx^n$.

To say that y varies directly as x^2, says that $y = kx^2$. It does not determine the factor of proportionality **k**. If however, one pair of values of x and y are known, k can be determined. Thus, if $y = 27$ when $x = 3$, then $27 = k3^2$, or $27 = 9k$ and $k = 3$. $y = 3x^2$, gives the exact relationship between x and y. From $y = 3x^2$ one can now deduce additional pairs of values by assigning values to x and computing the corresponding values of y.

Definition 3: A variable y is said to vary **inversely** as x, if $xy = k$ or $y = k/x$.

Definition 4: If three variables x, y, and z are so related that $z = kxy$, then z is said to vary **jointly** as x and y.

Definition 5: If three variables are so related that $z = k\frac{y}{x}$, then z is said to vary **directly** as y and **inversely** as x.

Illustration 1.

The intensity I of illumination varies inversely as the square of the distance d from the source of light. If the intensity at a distance of 2 feet is 100 candle power, what is the intensity at a distance of 3 feet?

Solution: First one must state the variation as $I = k/d^2$. Since $I = 100$ when $d = 2$, $100 = k/4$. This gives $k = 400$ and $I = 400/d^2$. To find I when $d = 3$, one has $I = 400/9 = 44\frac{4}{9}$.

Illustration 2.

The safe load L for a horizontal beam supported at both ends varies jointly as the breadth b and the square of the depth d, and inversely as the length l between supports. If a 2 by 6 inch beam 10 feet long safely supports 1000 pounds, what is the safe load for a 2 by 4 inch beam 12 feet long?

Solution: By the statement of the law of variation, $L = k\frac{bd^2}{l}$. For the given data $1000 = \frac{k \cdot 2 \cdot 36}{10}$. This gives $k = \frac{1,250}{9}$ so that the complete variation formula is

$$L = \frac{1,250}{9}\,\frac{bd^2}{l}.$$

For the 12 foot beam, $L = \frac{1,250}{9}\,\frac{2(16)}{12} = \frac{10,000}{27} = 370\frac{10}{27}$.

48. Summary of Procedure.

Four steps are essential in the solution of problems dealing with variation.

a) Write an equation representing the type of variations as given by the problem.

b) Use the data of the problem to determine the value of **k**, the factor of proportionality.

c) Write the explicit formula for the variation, using the computed value of **k**.

d) Find additional value or values using this formula.

Chapter VII

GRAPHS AND FUNCTIONS

49. Introduction.

In the preceding chapter we saw that there are certain relationships in the form of equations expressing variation, which have the property that from them sets of related values may be computed by assigning values to one or more of the variables involved. In the present chapter we shall examine the variation between variable quantities from a graphical point of view.

50. Charts of Data and Graphs.

Oftentimes a pictorial account of data presents facts in a form more easily comprehended than is usually possible from a mere listing of the data. Following is an example.

During the month of July, 1950, the approximate number of workers on the pay rolls of an organization were as follows:

July 3—1,611,000
July 10—1,619,000
July 17—1,659,000
July 24—1,689,000
July 31—1,700,000.

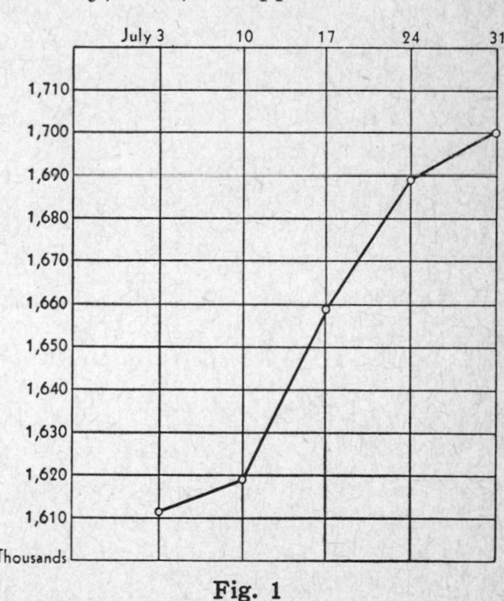

Fig. 1

This same information can be displayed graphically as shown in the adjoining graph, Fig. 1. This graph shows at a glance the period of greatest increase (July 10-17), as well as the fact that during the first and last weeks the increase was approximately the same.

Sometimes data are represented in the form of a "bar" graph. Consider, for example, the following data.

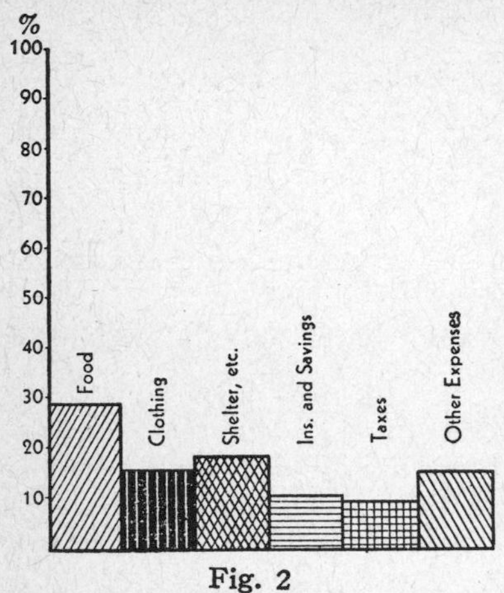

Fig. 2

In a certain household, the expenses may be classified as follows: Foods, 29%, Clothing, 16%, Shelter, Heat, and Light, 18%, Insurance and Savings, 11%, Taxes, 10%, All others (Auto, Health, Recreation, Benevolence, etc.), 16%. These data are displayed on a bar type graph in Fig. 2.

This graphical display does not show the relative distribution so well as does a circular arrangement in which the whole circle represents 100%.

For the circular chart see Fig. 3.

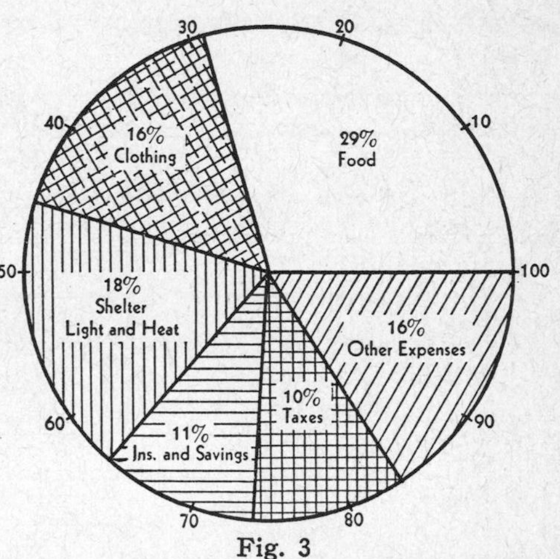

Fig. 3

Numerous other graphical schemes may be used,* but in connection with a course in algebra we shall ordinarily use a rectangular coordinate system and draw the graphical representation of equations.

51. A Coordinate System.

We consider now a device which will enable us to picture the relationship between number pairs and the points of the plane. In order to do this, we must make the following assumptions:

* See: *College Outline Series: An Outline of Statistical Methods*, Arkin and Colton, Chap. XVIII. (3rd ed.).

1. We have given two straight lines X X′ and Y Y′, perpendicular to each other, and intersecting in a point O.

2. We must choose a convenient distance which will be considered the **unit** of measure.

The two lines are called the **x-axis** and **y-axis,** respectively. (See Fig. 4.) Their point of intersection is called the **origin.** We

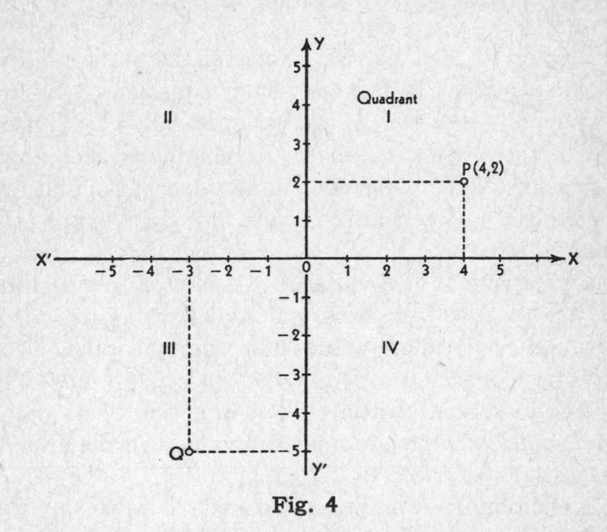

Fig. 4

further agree that successive unit distances will be laid off from the origin along each axis, and thus a scale of measure is established.

The positive values are to the right of the origin along the x-axis, and above the origin along the y-axis.

Negative values are laid off to the left along the x-axis, and below the origin along the y-axis.

The plane now is divided into four regions called **quadrants,** usually numbered as shown.

Any point, such as P in the figure, can be designated by giving values of x and y, called the x-coordinate (or **abscissa**) and the y-coordinate (or **ordinate**) of the point. In the figure, P is designated by $x = 4$, $y = 2$, usually written (4, 2). A point is designated by a number pair (x, y). Note that the x value is always given first, then the y value. Also in the figure, the point Q has the coordinates $(-3, -5)$.

By means of this coordinate system we have set up a correspondence such that for every point of the plane there is a particular number pair (x, y) which describes its location. Conversely, to every number pair (x, y) there corresponds a position in the plane. This is called a **rectangular coordinate system.**

52. Functions.

In Chapter VI, Sec. 47, we discussed the variation given by the equation $y = 2x$. In this particular equation, x and y are so related to one another that if we assign a value to x, a definite value of y is thereby determined. Furthermore, each and every value assigned to x definitely determines a corresponding value of y. This idea may be generalized as is now done in the following definition.

Definition 1: If a variable y is related to a variable x in such a way that each assignment of a value to x definitely determines one or more values of y, then y is called a **function** of x.

Thus the area of a circle is a function of its radius, for $A = \pi r^2$, and each assignment of a value to r definitely determines a value of the area.

Definition 2: The variable to which values are assigned is called the **independent** variable. The variable whose value is thereby determined is called the **dependent** variable.

Rather obvious extensions of these ideas can be made to situations involving three or more variable quantities. Thus $z = 4xy$ expresses the fact that z is a variable depending upon the two independent variables x and y. We say that z is a function of both x and y.

53. Functional Notation.

Rather than use words and sentences to express the definition which we gave for "function," we usually write $y = f(x)$. This is read, "y is a function of x."

Suppose that y is some particular function of x, say $x^2 - 5x + 6$. This may be written $y = f(x) = x^2 - 5x + 6$. When x equals zero, y has the value 6. When $x = 2$, then $y = 0$. When $x = -3$, then $y = 30$. If we understand that f(2) represents the value of this particular function when x has the value 2, then we may use our notation to express this fact. Thus, if

$$f(x) = y = x^2 - 5x + 6,$$

then $f(2) = y = 2^2 - 5 \cdot 2 + 6 = 0.$

Furthermore $f(0) = 6$

$$f(-3) = 30.$$

If $z = x^2 - 3xy + x^3y + y^2$, we may write

$$z = f(x, y) = x^2 - 3xy + x^3y + y^2,$$

and $z = f(0, 2) = 4,$

also $z = f(-1, 2) = 9.$

Whenever two or more functions are used in the same problem, one usually represents them by different functional symbols. Thus, if there are two functions $y = x^2 - 5x + 6$, and $y = x^3 - 4$, we may represent the first function by $f(x)$, the second by $F(x)$, in order to distinguish between them. An alternative notation would be to represent the first function by $f(x)$ and the second by $\phi(x)$. (ϕ is the corresponding Greek letter **phi.**) Still another representation would be $f_1(x)$ and $f_2(x)$, where the subscripts 1 and 2 identify the particular functions.

Example 1: Express in functional notation the fact that the area (A) of a triangle is a function of its base (b) and height (h).

Solution: We could write simply $A = f(b, h)$. But from our knowledge of plane geometry we know that the area of a triangle equals one half the product of the base by the height, so that $A = \frac{1}{2}bh$.

Example 2: If $f(x) = 3x^2 - 5$ and $F(x) = x + 4$, find the values of $f(2) - F(1)$ and also $\frac{f(-1)}{F(2)}$.

Solution: $f(2) = 3 \cdot 2^2 - 5 = 7$ and $F(1) = 1 + 4 = 5.$

Therefore $f(2) - F(1) = 7 - 5 = 2.$

Also, $f(-1) = 3(-1)^2 - 5 = -2$ and

$$F(2) = 2 + 4 = 6.$$

Therefore $\frac{f(-1)}{F(2)} = \frac{-2}{6} = -\frac{1}{3}.$

54. Graphical Representation of Functions.

Consider the equation $y = f(x) = 2x$, and compute a chart of values for $f(x)$ as shown below.

x	-3	-2	-1	0	1	2	3
$y = f(x)$	-6	-4	-2	0	2	4	6

If the pairs of values x and $f(x)$, or simply (x, y), are interpreted as points, then we have a set of pairs of values which we may plot as shown in Fig. 5. Other values may also be computed and plotted, but this we shall not do. However, it is important for us to realize that points may be obtained for as many values as one pleases between say the values $x = 1$ and $x = 2$. In analytic geometry, it is shown that all such points for an equation of the first degree in two variables lie on a straight line. Consequently, if we join the points found above by a line, we say that we have drawn the graph of the function. Sometimes we say that we have drawn the locus of the equation. By definition, the **locus** consists of the totality of those, and only those, points whose coordinates satisfy the given equation.

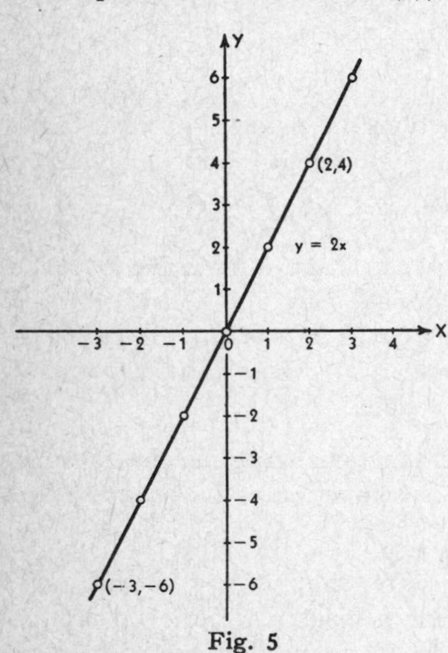

Fig. 5

55. Aids in Graphing.

The graph of any function may be obtained by simply plotting points. These points may be determined by substituting values of x and computing $f(x)$ from the given functions. The pairs of values thus found may be plotted on a coordinate system and then joined by a smooth curve. In general, such a procedure is tedious, so that further facts aside from points of the curve are very helpful. Consequently one usually proceeds to analyze a given function from the standpoint of intercepts, symmetry, extent and excluded values, and asymptotes. These special terms will now be considered separately and each will be illustrated.

(A) **Intercepts.** The **intercepts** of a curve are the points at which the curve cuts the axes. Since all points on the x-axis have y equal to zero, the x-intercepts may be obtained by setting $y = 0$ in the equation of the curve. The y-intercepts are similarly determined by setting $x = 0$.

(B) **Symmetry.** The points (3, 4) and (3, −4) are symmetrically located with respect to the x-axis. The points (2, 5) and (−2, 5) are symmetric with respect to the y-axis. The points (5, 3) and (−5, −3) are symmetric with respect to the origin.

In general a curve is symmetric with respect to the x-axis if the points (x, y) and (x, −y) both satisfy its equation. A curve is symmetric with respect to the y-axis if (x, y) and (−x, y) both satisfy its equation. A curve is symmetric with respect to the origin if (x, y) and (−x, −y) both satisfy its equation.

(C) **Extent and Excluded Values.** Sometimes a function has no real value for a given value of the independent variable. Such a value for the independent variable is called an **excluded value.** If, for all the values from $x = a$ to $x = b$, there are no corresponding values for y, then the region of the plane between a and b is called an **excluded region.** The presence of such regions for some functions naturally limits the extent of the graph of the function. These excluded values usually arise from the fact that for a given value of x, the value of y involves the square root of a negative number. This would mean that the value of y is not real.

(D) **Asymptotes.** For our purpose here, we shall set up a restricted definition of an asymptote. Suppose that a given straight line is so situated with respect to a curve, that a point moving along the curve always approaches the line indefinitely (yet never reaches it) and at the same time the point is receding from the origin. Such a line is called an **asymptote** of the curve.* We shall be interested only in the horizontal and vertical asymptotes of curves at this time.

---

* The general definition of asymptote depends upon the limiting process as used in calculus and generally occurs in a course in the calculus.

Illustration 1.

Discuss and draw the graph of $y = x^2$. Substituting $x = -x$ in the equation gives $y = (-x)^2 = x^2$, thus leaving it unchanged; therefore the graph is symmetrical with respect to the y-axis. If one replaces y by $-y$, the equation is changed into $-y = x^2$, and consequently there is no symmetry with respect to the x-axis. The x and y intercepts are at the origin, since $x = 0$ gives $y = 0$ and conversely.

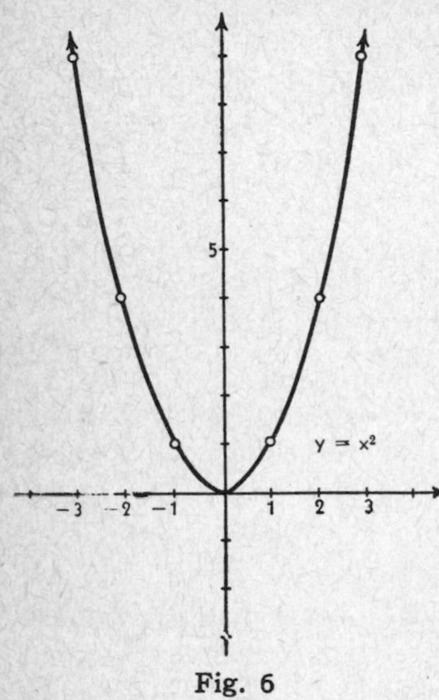

Fig. 6

The value of y is definitely determined for every positive or negative value of x and is always positive. The y value increases with an increase in the value of x. Since no real value of x can make y negative, none of the curve lies below the x-axis, and that region is excluded.

By examining y for a few values of x, and making use of the facts shown above, especially the symmetry aspect, one obtains a graph as shown in Fig. 6.

x	0	1	2	3
y	0	1	4	9

Illustration 2.

Discuss and draw the graph of $y = x^3$. The curve passes through the origin, since $x = 0$ gives $y = 0$. The tests for symmetry show that it is not symmetric with respect to either axis. However, $x = -x$ and $y = -y$ gives $-y = (-x)^3 = -x^3$, which is equivalent to $y = x^3$, and the curve is symmetric with respect to the origin. Since y is positive if x is positive, and y is negative if x is negative, the entire graph lies in the first and third quadrants. The second and fourth quadrants constitute the excluded regions. By finding coordinates of a few points of the curve for first quadrant values, and making use of the symmetry with respect to the origin, one obtains the curve which appears in Fig. 7.

x	0	1	2	3	
y	0	1	8	27	

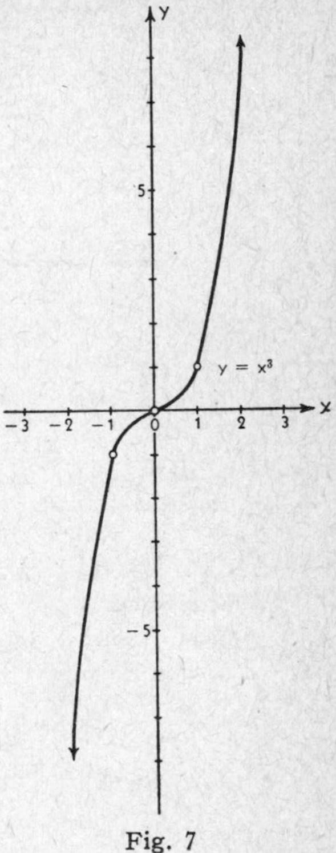

Fig. 7

Illustration 3.

Discuss and draw the graph of $xy = 1$. This may be written $y = 1/x$. Applying the tests for symmetry, we find it to be symmetric with respect to the origin. If x is positive, so is y, and this coupled with the fact that the function is symmetric with respect to the origin shows us that the second and fourth quadrants are excluded regions. When one assigns $x = 0$ to ascertain the y-intercept, the y value is not defined since division by zero is undefined. (See Sec. 6.) Consequently, we examine what happens when x approaches zero, as shown in the following chart.

x	1	.1	.01	.001	.00001	etc.
y	1	10	100	1000	100000	

We see that as x gets continually smaller (yet not actually taking the value zero), the values of y increase indefinitely. This means that the y-axis is a vertical asymptote. If we write the equation in the form $x = 1/y$, we see that the x-axis is a horizontal asymptote. The graph may now be obtained by finding a few more points of the curve as now shown.

x	10	5	1	$\frac{1}{2}$	$\frac{1}{10}$	
y	$\frac{1}{10}$	$\frac{1}{5}$	1	2	10	

The curve is as shown in Fig. 8.

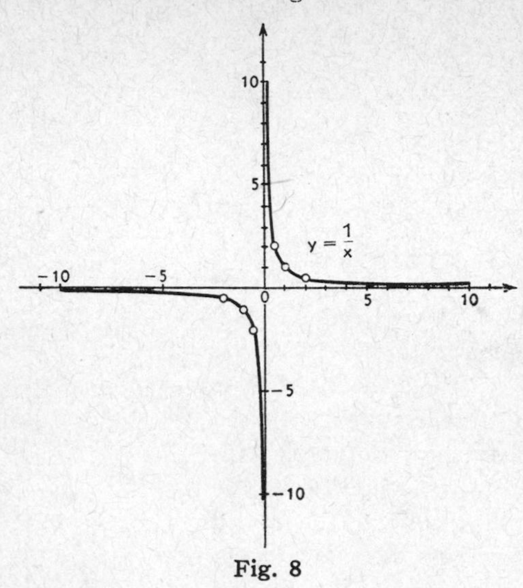

Fig. 8

Illustration 4.

Discuss and draw the graph of $x^2 + y^2 = 25$. This curve is symmetrical with respect to both axes and to the origin. If $x = 0$, $y^2 = 25$, so that $y = \pm 5$ and the y-intercepts are $+5$ and -5. Likewise its x-intercepts are ± 5. If we write the equation as $y^2 = 25 - x^2$, so that $y = \pm\sqrt{25 - x^2}$, we see that no value of x which is greater than 5 will give a real value to y. Since the sum of the squares of x and y is always equal to 25, the locus is a circle of radius 5 and center at the origin. See Fig. 9.

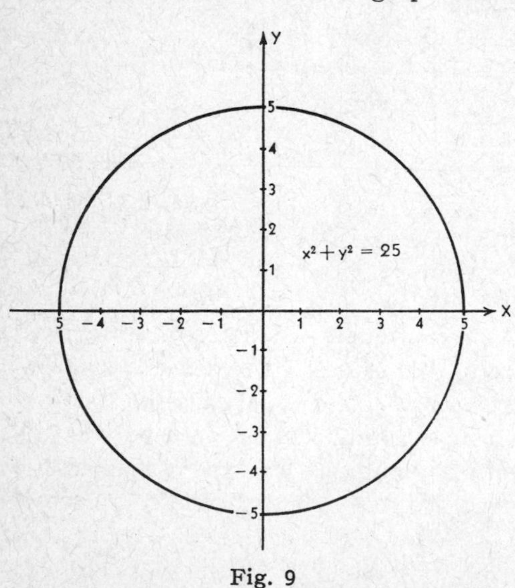

Fig. 9

Illustration 5.

Discuss and sketch the curve $y = \frac{x + 3}{x^2 - 4}$. If $x = 0$, $y = -\frac{3}{4}$ and this is the only y-intercept. If $y = 0$, $x = -3$ and this is the only x-intercept. The curve is not symmetric with respect to either axis or the origin.
There are two values which will cause the denominator to vanish, namely $x = 2$ and $x = -2$. Consequently the lines $x = 2$ and $x = -2$ are each vertical asymptotes. The given equation may be written $yx^2 - x - (4y + 3) = 0$. If this is solved for x by the quadratic formula (see Chapter IX) one has $x = \frac{1 \pm \sqrt{1 + 12y + 16y^2}}{2y}$. The denominator in this case vanishes if $y = 0$, so that the x-axis is a horizontal asymptote. Furthermore, y is negative for every value of x between -2 and $+2$. It is positive for x greater than 2.

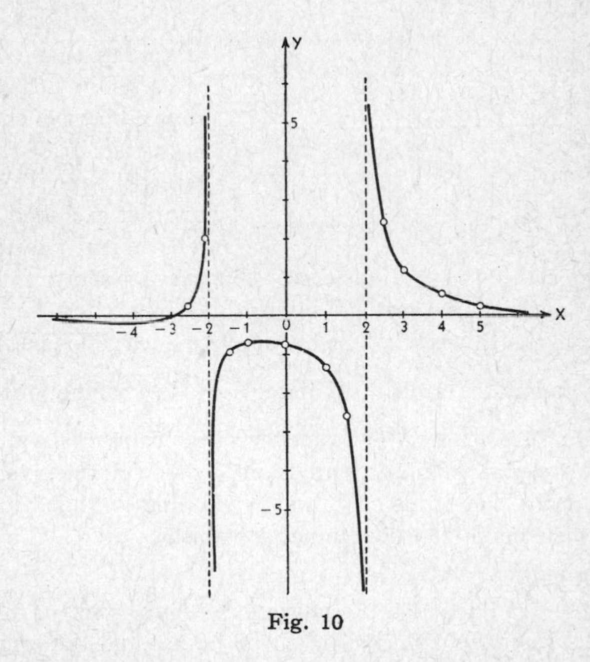

Fig. 10

These facts along with a few computed values give us a curve as shown in Fig. 10.

Illustration 6.

Discuss and sketch the curve $y^2 = \frac{x+3}{x^2-4}$. Note that the only difference between this and the equation of Illustration 5 is the y^2 instead of y. This change alters the curve considerably. The x-intercept as before is -3. But when $x = 0$, then $y^2 =$ a negative number so that there is no real y-intercept. Besides, any value of x between -2 and $+2$ gives y^2 a negative value so that this becomes an excluded region. Every value of x contributes 2 values of y, and the curve is symmetric with respect to the x-axis. If x is less than -3, then y^2 is again negative, so that no part of the curve is to the left of the x-intercept, and we have another excluded region. The asymptotes are the same as in the previous case. The curve is shown in Fig. 11.

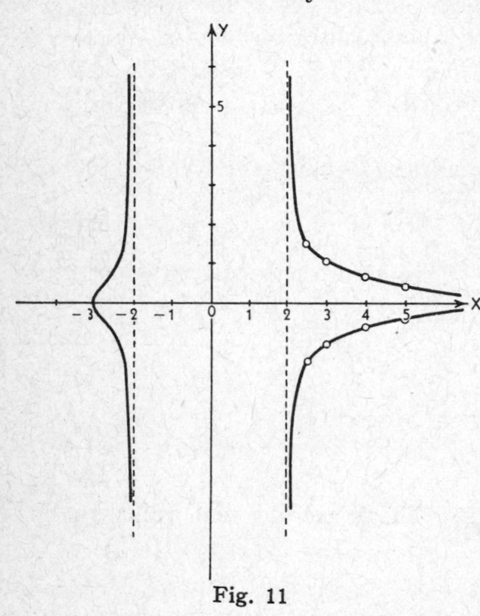

Fig. 11

In general the vertical asymptotes to a curve are found from the values of x which will make a variable factor in the denominator vanish. Thus $y = \frac{x}{x^2+4}$ has no vertical asymptotes, since $x^2 + 4$ is never zero for any real value of x. Horizontal asymptotes are found by solving the given equation for x in terms of y, and proceeding to find values of y which make the denominator vanish.

Illustration 7.

Discuss and sketch the curve $4x^2 + 9y^2 = 36$. This curve is symmetric with respect to both axes and the origin. If $x = 0$, $y = \pm 2$, giving the y-intercepts. If $y = 0$, $x = \pm 3$, giving the x-intercepts.

Solving for y we have $y = \pm\frac{2}{3}\sqrt{9 - x^2}$, and thus no value of x greater than 3 will give a real value for y. Solving for x we have $x = \pm\frac{3}{2}\sqrt{4 - y^2}$, and thus no value of y greater than 2 will give a real value for x. Therefore the curve must lie entirely within a rectangle bounded by the lines $x = 3$, $x = -3$, $y = 2$, $y = -2$. This curve is an ellipse, as shown by the graph in Fig. 12. In general, every curve of the type $Ax^2 + By^2 = C$ where A, B, and C are positive quantities, is an **ellipse.**

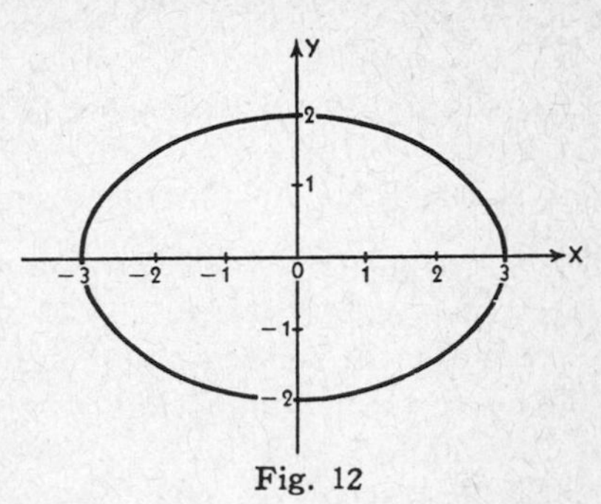

Fig. 12

Illustration 8.

Discuss and sketch the curve $4x^2 - 9y^2 = 36$. This curve is symmetric with respect to both axes and the origin. The x-intercepts are ± 3. If $x = 0$, y^2 is negative so that there are no real y-intercepts. Solving the equation for y we have $y = \pm\frac{2}{3}\sqrt{x^2 - 9}$. There are no real values of y for values of x between -3 and $+3$, and this is therefore an excluded region.

If x is greater than 3, then y increases with increasing values of x. The curve is a hyperbola and is shown in Fig. 13.

In general, $Ax^2 - By^2 = C$, where A, B, and C are positive, is a **hyperbola.**

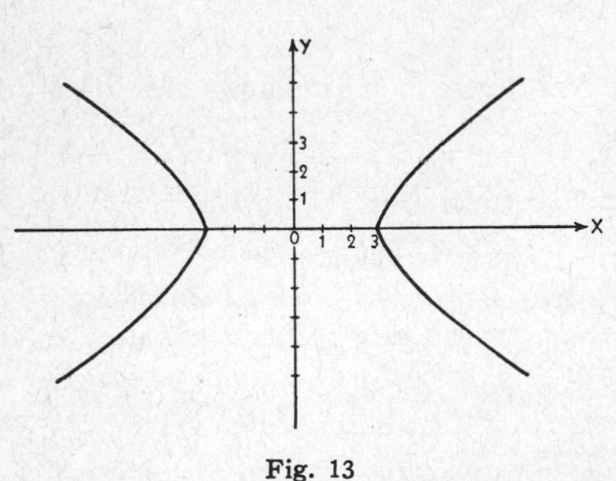

Fig. 13

Graphical representation of other functions are to be found on the following pages: 61, 62, 63, 64, 70, 71, 74, 79, 80, 81, 85, 117, 118, 135, 170, 172, 173.

Chapter VIII

LINEAR EQUATIONS

56. Linear Equations in One Unknown.

The simplest possible type of equation involves one unknown and is of the first degree. An example would be $2x - 6 = 0$. The solution of this equation is obtained by adding 6 to both sides of the equation, yielding $2x = 6$; then dividing both sides by 2, yielding $x = 3$.

In general, such an equation would be of the form $ax + b = 0$, and its solution is $x = -b/a$.

If the equation is not in this general form, it can always be reduced to it.

Illustration.

Solve $3x - 2 = 5(x + 4)$.

This may be written:

$$\begin{aligned} 3x - 2 &= 5x + 20 \\ 3x - 5x &= 20 + 2 \\ -2x &= 22 \\ x &= -11. \end{aligned}$$

An equation of the first degree in one unknown always has one and only one solution.

57. A Linear Equation in Two Unknowns.

Such an equation involving two unknowns, say x and y, would be of the type $2x - 3y = 7$. There are an indefinitely large number of pairs of values which satisfy this equation. One such pair would be $x = 2$, $y = -1$. Any other solution may be found by first expressing y as a function of x as in Chapter VII. This gives $y = \frac{2x - 7}{3}$. If now a value is assigned to x, the corresponding value of y may be computed.

Any number of such pairs (or solutions) may be found, but since it can be proved, usually in analytic geometry, that every equation of the first degree in two unknowns has a graph which is a straight line, only two such solutions are necessary to obtain the graph. It is useful to compute three such pairs, generally to assist in avoiding errors. For, if all three values lie on the same line one can be reasonably sure that no error of computation has been made.

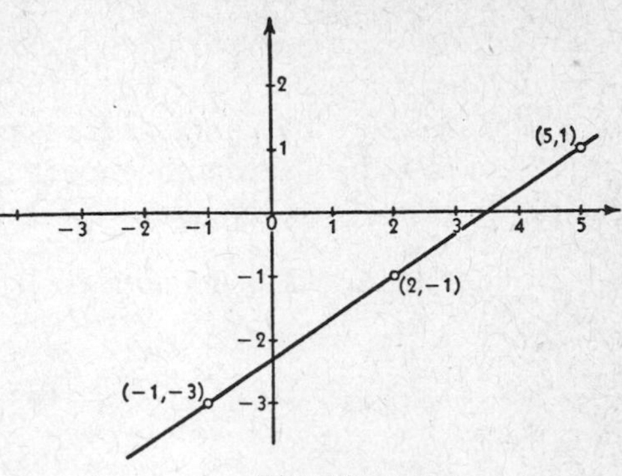

Fig. 14

Below is a chart of values and the graph of the equation $2x - 3y = 7$. (See Fig. 14.)

x	−1	2	5
y	−3	−1	1

The general form of the equation can be written as $ax + by = c$, where a, b, and c are numerical values (constants) called the **coefficients.**

58. System of Two Linear Equations (2 Unknowns).

Since every linear equation in two unknowns has a graph which is a straight line, one might expect that if two such equations are graphed, they might have points in common.

In fact, one of three situations must be true:

a) The graphs have only **one** point in common (intersecting lines).

b) The graphs have **no** points in common (the lines are parallel).

c) The graphs have **all** points in common (the lines coincide).

If case (a) holds, the equations are said to have a **simultaneous** solution. That is, there is a pair of values (x, y) which satisfies both equations. Ordinarily, this common value could be only approximated by graphical methods, since graphs can be drawn only with a limited degree of accuracy. Therefore, we shall use the graphical solution as an interpretation of the meaning of the algebraic solution. We proceed now to discuss algebraic means for obtaining the simultaneous solution.

The three classifications as given above are also referred to by saying that:

In case (a) the equations are **consistent,**
In case (b) the equations are **inconsistent,**
In case (c) the equations are **equivalent** or **dependent.**

59. Solution by Addition or Subtraction.

Consider the two equations:

$$3x - 4y = 7 \qquad (1)$$
$$x + 6y = 6. \qquad (2)$$

Repeating the first equation, and writing an equivalent expression for the second, obtained by multiplying each term by 3 in order to make the coefficients of x in the two equations equal, one has:

$$3x - 4y = 7 \qquad (3)$$
$$\text{and } 3x + 18y = 18. \qquad (4)$$

Subtracting equation (4) from equation (3), we have:

$$-22y = -11 \qquad \text{or}$$
$$y = \frac{-11}{-22} = \tfrac{1}{2}.$$

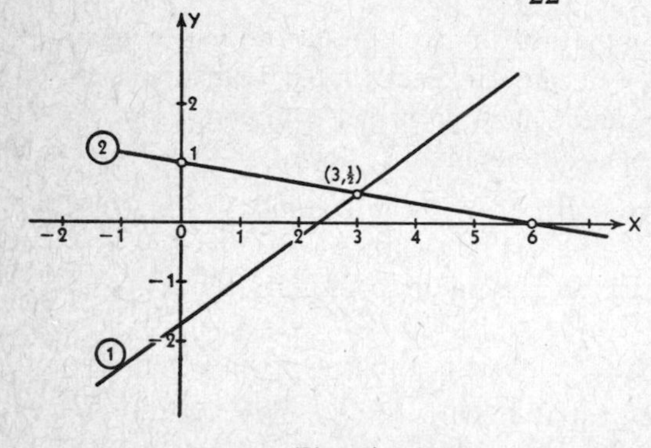

Fig. 15

Substituting $y = \frac{1}{2}$ in equation (1) gives

$$3x - 2 = 7 \text{ or}$$
$$3x = 9,\ x = 3.$$

This solution is usually written $(3, \frac{1}{2})$, and represents the point which the two lines have in common. The graphical interpretation of this solution is shown in Fig. 15, where the lines are labeled (1) and (2) to correspond to the equations as given.

60. Solution by Substitution.

Suppose now that we solve the equations given in the preceding section as follows:

From the second equation we have:

$$6y = 6 - x \text{ or } y = \frac{6 - x}{6}.$$

Substituting this value of y in the first equation gives

$$3x - 4\,\frac{6 - x}{6} = 7.$$

Multiply each term by 6 to rid the equation of fractions, and we have: $18x - 24 + 4x = 42$.

Collect terms, and one has: $22x = 66$ or $x = 3$.

The value $x = 3$ may now be substituted in either equation, giving $y = \frac{1}{2}$, as was found before.

61. Inconsistent and Dependent Cases.

Illustration 1.

The equations $2x + 3y = 6$ and $2x + 3y = 12$ represent parallel lines, as shown by Fig. 16. If we try to solve for

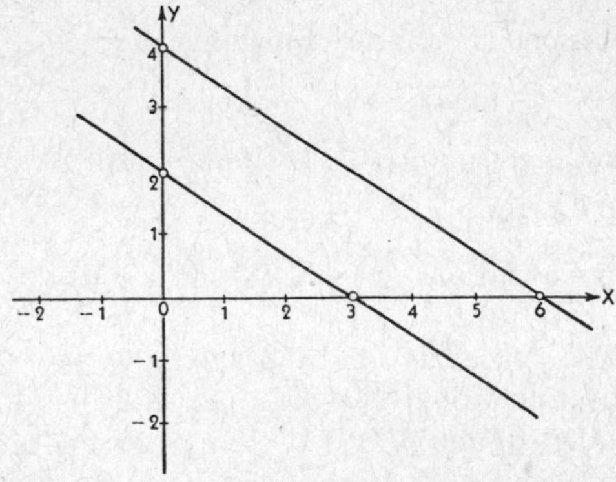

Fig. 16

x from this given pair we arrive at $0 = -6$, an inconsistency.

Illustration 2.

The equations $2x + 3y = 6$ and $\frac{2}{3}x + y = 2$ represent the same line. In fact, if the second equation is multiplied by 3, the result is the first equation. Trying to solve for x or y, leads to $0 = 0$. Any pair of values of x and y which satisfies one of the equations satisfies the other also. See Fig. 17.

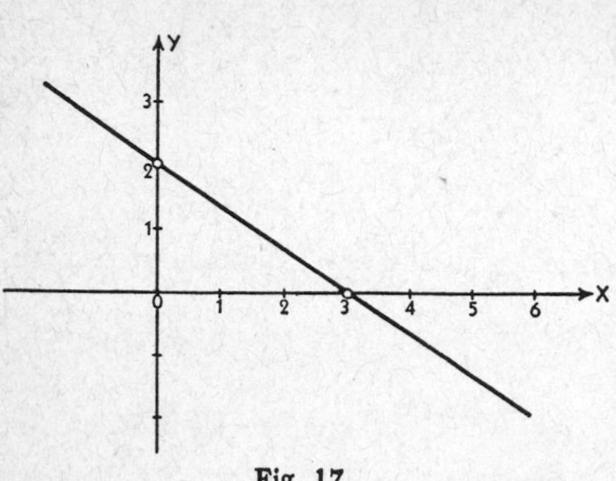

Fig. 17

62. Systems of Three Linear Equations (3 Unknowns).

A system of three linear equations in three unknowns may be solved by first eliminating one of the unknowns and thereby reducing the problem to that of solving a system of two equations in two unknowns. The method will be illustrated by an example.

Solve the system:

$$x + y + z = 2. \qquad (1)$$
$$2x - 2y - z = 2. \qquad (2)$$
$$x + 2y - z = -3. \qquad (3)$$

Add equations (1) and (2) to eliminate z:

$$3x - y = 4. \qquad (4)$$

Add equations (1) and (3) to eliminate z:

$$2x + 3y = -1. \qquad (5)$$

Now solve equations (4) and (5) for x and y by eliminating, say y.

To do this: Multiply (4) by 3, $9x - 3y = 12.$ (6)

Repeat equation (5) $2x + 3y = -1$ (7)

Adding (6) and (7) $11x = 11$

or $x = 1.$

Substitute $x = 1$ in either equation (4) or (5) obtaining $y = -1$. Using these values of x and y in any of the three given equations, obtain $z = 2$.

The solution is usually written $(1, -1, 2)$ in which the values of the unknowns appear in the same order as they appeared in the given system of equations. The method is general and will apply to a system of n equations in n unknowns.

63. Other Methods of Solution.

For those who wish to solve systems of linear equations by means of determinants, Chapter XX may be taken up at this time.

Chapter IX

THE QUADRATIC EQUATION IN ONE UNKNOWN

64. Introduction.

So far, we have been considering the solution of linear equations. We now begin a study of equations of higher degree than the first, by studying the quadratic equation in one unknown.

Definition: A second degree (quadratic) equation in one unknown x, which can be expressed in the form $ax^2 + bx + c = 0$, will be referred to in this chapter simply as a quadratic equation.

The coefficients, a, b, and c, represent constants, and we shall confine our attention to the case where these constants represent real numbers. Specific examples of quadratic equations would be:

$2x^2 - 5x + 7 = 0$ Here $a = 2, b = -5, c = 7$.
$x^2 - 9 = 0$ Here $a = 1, b = 0, c = -9$.
$3x^2 + 2x = 0$ Here $a = 3, b = 2, c = 0$.

We should remark at this point that we do not have a case where $a = 0$, since then the equation would have no term of the second degree and would then be linear in type. Furthermore, the equation cannot have terms of higher degree than the second.

We adopt the form $ax^2 + bx + c = 0$ as representing the type form of the quadratic equation because any equation of the second degree in one unknown can be reduced to this form.

Consider the equation

$$(2x - 3)^2 + (x - 1)(x + 2) - 4x + 5 = 2.$$

If we simplify this equation by performing the indicated operations and collecting terms, we have:

$$4x^2 - 12x + 9 + x^2 + x - 2 - 4x + 5 - 2 = 0 \quad \text{or}$$
$5x^2 - 15x + 10 = 0$ or dividing out the factor 5,
$x^2 - 3x + 2 = 0$, which is now in type form with $a = 1, b = -3, c = 2$.

65. Incomplete Quadratics and Their Solutions.

Definition: If all of the terms of the type form are not present, the quadratic is said to be **incomplete.**

Example 1: Consider the incomplete quadratic equation $x^2 - 2x = 0$. This equation can be solved by means of factoring as follows. Write the equation as $x(x - 2) = 0$. Then one of the factors of this product must be zero. If we set the first equal to zero, we have $x = 0$ as one solution. If we set the second equal to zero, we have $x - 2 = 0$, or $x = 2$ as the other solution.

Example 2: Consider now the equation $x^2 - 9 = 0$. This can be solved simply by extracting the square root of both sides of the equation after writing it in the form $x^2 = 9$.

This gives $x = \pm 3$, meaning that we have two solutions, $x = 3$ and $x = -3$. This could have been obtained by factoring the given expression into $(x - 3)(x + 3) = 0$.

Example 3: Consider the equation $x^2 + 9 = 0$. If we recall the quadratic forms with their factors as considered in Sec. 16, we see that this simple quadratic equation cannot be solved by means of facts there presented. Indeed this simple equation leads us to quite a novel situation. One can, however, proceed as follows.

Write the equation as $x^2 = -9$. Then extract the square root of both sides of this equation obtaining $x = \pm\sqrt{-9}$. But how can we extract the square root of a negative number? The result can be neither $+3$ nor -3, for the square must be -9; and neither of these numbers when squared gives -9, but each gives $+9$. This leads us to discuss the situation in the next paragraph.

66. Square Roots of Negative Numbers.

We shall see presently that the square root of a negative number is a special type of number. Like the numbers with which we are already familiar, these new numbers behave properly under the fundamental operations and also lead us to very interesting properties and results. Before discussing them further we must consider a few definitions.

Definition 1: A number which is the square root of a negative number or which involves square roots of negative numbers is called an **imaginary number.**

Definition 2: We designate $\sqrt{-1}$ by i (*i.e.*, $i = \sqrt{-1}$) and call this number the **imaginary unit.** Its role is similar to the number 1 in the ordinary number system.

Definition 3: A number of the form $a + bi$ where a and b are real numbers is called a **complex number.**

Definition 4: The two complex numbers $a + bi$ and $a - bi$ are called **conjugate** complex numbers, and each is the conjugate of the other.

We have seen in Sec. 38 that a number such as $\sqrt{12}$ can be simplified to the form $\sqrt{(4)(3)} = 2\sqrt{3}$. In a similar way we can write an imaginary number, such as $\sqrt{-4}$, as $\sqrt{4(-1)} = 2\sqrt{-1} = 2i$. In fact, any number $\sqrt{-a}$ can be written $\sqrt{a(-1)} = \sqrt{a}\sqrt{-1} = \sqrt{a}\cdot i$.

Now since i is the square root of -1, we have also that $i^2 = -1$; that is, it is a number such that its square is the negative unit.

In order to show how one performs the fundamental operations with imaginary numbers, we list the following examples:

$$\sqrt{-4} + \sqrt{-9} = 2i + 3i = (2 + 3)i = 5i.$$

$$\sqrt{-5} + \sqrt{-4} - \sqrt{-9} = \sqrt{5}i + 2i - 3i = (\sqrt{5} + 2 - 3)i$$

$$= (\sqrt{5} - 1)i.$$

$$(\sqrt{-9})(\sqrt{-2}) = (3i)(\sqrt{2}i) = 3\sqrt{2}i^2 = 3\sqrt{2}(-1) = -3\sqrt{2}.$$

In forming products of imaginary numbers we must always write them in the form ki and mi and form the product $kmi^2 = km(-1) = -km$.

As examples of complex numbers we have:

$2 + 5i$, $-3 + \sqrt{2}i$, $\frac{1}{2} - \sqrt{\frac{3}{2}}i$, etc.

The conjugate of $2 + 5i$ is $2 - 5i$.

The sum of $(2 + 5i) + (2 - 5i)$ is 4 since $5i - 5i = 0$.

The sum of $(2 + 5i) + (7 - 6i)$ is $9 - i$.

The product $(2 + 5i)(2 - 5i)$ is evaluated by the rule given in Sec. 16 and is the product of the sum and difference of two quantities and therefore equal to the square of the first minus the square of the second.

Therefore $(2 + 5i)(2 - 5i) = 2^2 - (5i)^2 = 4 - 25i^2 = 4 + 25 = 29$.

We now prove two theorems which are especially useful in connection with the theory of quadratic equations.

Theorem I. The sum of a conjugate pair of complex numbers is a real number.

Proof: Take $a + bi$ and $a - bi$ as the conjugate pair. Then the sum $a + bi + a - bi = 2a$, which is a real number since a and b are real numbers.

Theorem II. The product of a conjugate pair of complex numbers is a real positive number.

Proof: Take $a + bi$ and $a - bi$ as the conjugate pair. Then $(a + bi)(a - bi) = a^2 - b^2i^2 = a^2 + b^2$. Since a and b are real numbers, and since whether positive or negative their squares will be positive, the sum of two positive numbers is again a positive number.

A more complete discussion of complex numbers occurs in Chapter XVII, but what we have given here is sufficient for our needs at this time.

67. Various Methods for Solving Quadratic Equations.

We now consider in order, the four most common methods of solving a quadratic equation. These are as follows: a) Graphical Solution, b) Factoring, c) Completing the Square, and d) Solution by Formula.

a) Graphical Solution.

This method depends upon the ideas which we first met in the chapter on Graphs and Functions (Chapter VII). It is an approximate method in most instances, but should give us a good understanding of what we mean by **solving** a quadratic equation.

Consider the equation $x^2 - 5x + 6 = 0$. If we assign values to x we obtain various values for the entire expression. In other words, we may employ the function notation and write:

$$f(x) = x^2 - 5x + 6.$$

The following chart shows the values of x used and the corresponding values for the function.

x	−3	−2	−1	0	1	2	2.5	3	4	5	6	7
f(x)	30	20	12	6	2	0	−.25	0	2	6	12	20

If we plot these values on a coordinate system, measuring x along the x-axis and f(x) along the y-axis, we obtain a curve as appears in Fig. 18.

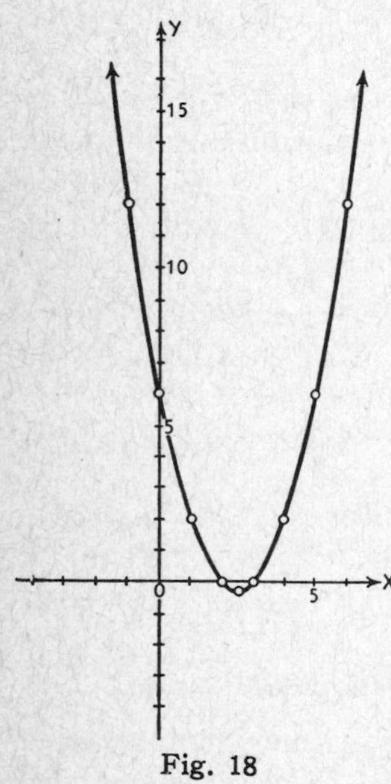

Fig. 18

The graph shows only some of the pairs of values in the chart. The student can easily verify that for values of x such as 9, 10, 11, etc., or $x = -4, -5, -6$, etc., the value of the function increases rapidly. In other words, the graph continues upward as is indicated by the arrows but at the same time no portion of it is a straight line. Such a curve as shown here is called a **parabola.** The curve crosses the x-axis at the points where $x = 2$ and $x = 3$. For these values of x, f(x) equals zero. In other words, $x = 2$ and $x = 3$ satisfy the equation $x^2 - 5x + 6 = 0$, and these two values are called the **roots** or **solutions** of the equation $x^2 - 5x + 6 = 0$.

Definition: Any value, or values, of x which satisfy an equation are called the **roots** (or **solutions,** or **zeros**) of the equation.

As a second example, consider the equation $-x^2 + 2x + 6 = 0$. Again, let $f(x) = -x^2 + 2x + 6$, and obtain the following chart by assigning values to x and computing the corresponding values of f(x).

x	−4	−3	−2	−1	0	1	2	3	4	5	6
f(x)	−18	−9	−2	3	6	7	6	3	−2	−9	-18

The graph in this case, Fig. 19, is inverted from the first example. The roots are not integral and appear to be **approximately,** $x = -1.5$ and $x = 3.7$. The exact values cannot be obtained from the graph. We shall see later just how to determine exactly the roots for this particular equation

Note. These two equations characterize several facts concerning quadratic equations in one unknown.

1. If the coefficient of the x^2 term is positive, the graph opens upward.
2. If the coefficient of the x^2 term is negative, the graph opens downward.
3. The graph of every equation $f(x) = ax^2 + bx + c$ is like the above curves, its size and position with reference to the coordinate axes being determined by the particular values of a, b, and c.

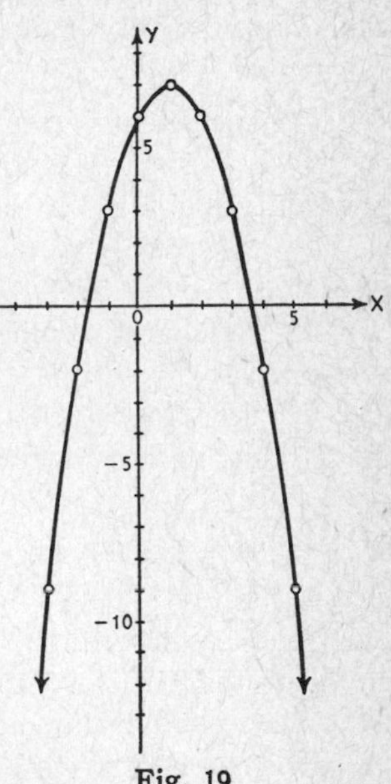

Fig. 19

b) Solution by Factoring.

Under many circumstances a quadratic equation can be solved by referring to the second degree forms of Sec. 16.

We illustrate the method by two examples.

Example 1: Given $x^2 - x - 12 = 0$.

This factors easily into $(x - 4)(x + 3) = 0$, and we employ the fact that if the product of two or more quantities is equal to zero, one or more of the factors must be zero. Consequently, $x - 4 = 0$ or $x + 3 = 0$, which gives two values for x, namely $x = 4$ and $x = -3$. Substitution of these values for x checks the fact that they are the two solutions.

Example 2: Given $2x^2 + 5x - 3 = 0$.

This may be written $(2x - 1)(x + 3) = 0$. Consequently $2x - 1 = 0$ or $x + 3 = 0$. These factors in turn yield the solutions $x = \frac{1}{2}$ and $x = -3$.

c) Solution by Completing a Square.

This method depends upon the fact that if one squares the

quantity $(x + k)$ one obtains $(x + k)^2 = x^2 + 2kx + k^2$, and one sees that the third term of the right hand expression is the square of half the coefficient of x in the second term. We employ this fact as follows:

Consider the equation $x^2 - 5x + 6 = 0$.

By transposing the 6 to the right hand side, we have:

$$x^2 - 5x = -6.$$

The coefficient of x is -5, half of this value is $-\frac{5}{2}$.

By squaring $-\frac{5}{2}$ and adding this value to each side of the equation, we obtain

$$x^2 - 5x + \tfrac{25}{4} = -6 + \tfrac{25}{4} = -\tfrac{24}{4} + \tfrac{25}{4} \quad \text{or}$$

$$x^2 - 5x + \tfrac{25}{4} = \tfrac{1}{4}.$$

The left hand member is now a perfect square and may be rewritten as:

$$(x - \tfrac{5}{2})^2 = (\tfrac{1}{2})^2.$$

Extracting the square root of both sides, we have:

$$x - \tfrac{5}{2} = \pm\tfrac{1}{2}.*$$

Hence $x = \frac{5}{2} \pm \frac{1}{2}$ yields two values for x,

$$x = \tfrac{5}{2} \pm \tfrac{1}{2}, \textit{ i.e.}, x = 3 \text{ or } 2.$$

Comparison with the first example under (a) of this section shows us that these solutions agree with those obtained before.

As a second example consider $-x^2 + 2x + 6 = 0$.

This can be written $-x^2 + 2x = -6$ or $x^2 - 2x = 6$.

Taking half the coefficient of x, squaring it and adding it to both sides,

$$x^2 - 2x + 1 = 6 + 1 = 7 \text{ or}$$

$$(x - 1)^2 = (\sqrt{7})^2 \quad \text{So that}$$

$$x - 1 = \pm\sqrt{7} \text{ and } x = 1 \pm \sqrt{7}$$

$$\textit{i.e.}, x = 1 + \sqrt{7} \text{ or } x = 1 - \sqrt{7}.$$

These two values of x are the exact values to which reference was made at the end of the discussion of the second graphical solution under case (a) of this section.

d) Solution by Formula.

By applying the method of section (c) to the general

* Note: In extracting the square root, the double sign might be placed on both sides, but to do so does not contribute any more solutions since $- = -$ is the same as $+ = +$. And $- = +$ is the same as $+ = -$.

equation $ax^2 + bx + c = 0$ one may develop a formula for solving a quadratic equation. Since by assumption **a**, the coefficient of x^2, was different from zero, we may divide both sides of the equation by **a** and obtain

$$x^2 + \frac{b}{a}x + c/a = 0.$$

Transposing c/a to the right hand side,

$$x^2 + \frac{b}{a}x = -c/a.$$

Completing the square by adding $b^2/4a^2$ to both sides:

$$x^2 + \frac{b}{a}x + b^2/4a^2 = b^2/4a^2 - c/a = \frac{b^2 - 4ac}{4a^2} \quad \text{or}$$

$$(x + b/2a)^2 = \frac{b^2 - 4ac}{4a^2}.$$

Extracting the square root of both sides,

$$x + b/2a = \frac{\pm\sqrt{b^2 - 4ac}}{2a}.$$ Hence we have the quadratic formula

$$\mathbf{x} = \frac{-\mathbf{b} \pm \sqrt{\mathbf{b}^2 - 4\mathbf{ac}}}{2\mathbf{a}}.$$

By using first the + and then the − sign before the radical we obtain the two roots of the quadratic equation.

Example: Solve $3x^2 - 5x - 2 = 0$.

Here $a = 3$, $b = -5$, and $c = -2$. Using the formula, and noting that $-b = -(-5) = +5$ and $-4ac = -4(3)(-2) = +24$, we have:

$$x = \frac{5 \pm \sqrt{25 + 24}}{6} = \frac{5 \pm \sqrt{49}}{6} = \frac{5 \pm 7}{6}$$

$$= 2 \text{ or } -\tfrac{1}{3}.$$

68. Theory Associated with a Quadratic Equation.

If a quadratic equation is represented by the form

$$ax^2 + bx + c = 0, \text{ then,}$$

1. The roots are given by the formula $x = \dfrac{-b \pm \sqrt{b^2 - 4ac}}{2a}$
2. The equation always has two roots, namely,

$$x = \frac{-b + \sqrt{b^2 - 4ac}}{2a} \quad \text{and} \quad x = \frac{-b - \sqrt{b^2 - 4ac}}{2a}$$

3. The quantity $b^2 - 4ac$ is defined as the **discriminant** of the quadratic equation because it distinguishes the kinds of roots which the quadratic equation has.

4. If the discriminant, $b^2 - 4ac$, equals zero, the equation has two equal roots. These values are each equal to $-b/2a$, as is seen from (2) above.

Graphically this means that the parabola is tangent to the x-axis at a point where $x = -b/2a$. (See Fig. 20.)

5. If the discriminant is a positive number, then the equation has two real and distinct roots whose values are given by (2) above.

Case 1. If $b^2 - 4ac$ is a perfect square, the roots are real, rational, and unequal.

Case 2. If $b^2 - 4ac$ is not a perfect square, the roots are real, irrational, and unequal.

In either case, the parabola crosses the x-axis at two distinct real points. (See Figs. 18, 19, 21.)

6. If the discriminant is negative, the roots given by (2) are conjugate complex numbers. This follows from the fact that we must take the square root of a negative number, and from Definition 4, Sec. 66 of this chapter. The graphical situation is that the parabola does not cut the x-axis. (See Fig. 22.)

Graphical Interpretation for $b^2 - 4ac \gtreqless 0$.

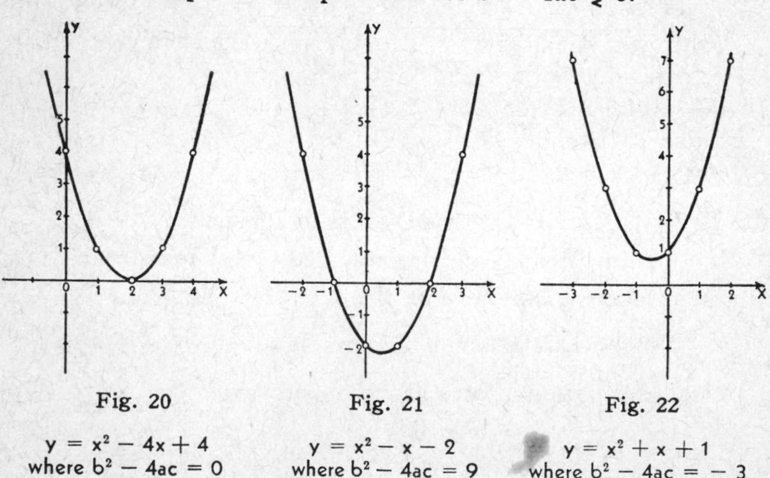

Fig. 20

$y = x^2 - 4x + 4$
where $b^2 - 4ac = 0$

Fig. 21

$y = x^2 - x - 2$
where $b^2 - 4ac = 9$

Fig. 22

$y = x^2 + x + 1$
where $b^2 - 4ac = -3$

7. The sum of the roots of a quadratic equation is equal to $-b/a$.

Proof: From (2) the roots are $x = \dfrac{-b + \sqrt{b^2 - 4ac}}{2a}$.

$$x = \frac{-b - \sqrt{b^2 - 4ac}}{2a}.$$

The sum of these roots is:

$$\frac{-b + \sqrt{b^2 - 4ac}}{2a} + \frac{-b - \sqrt{b^2 - 4ac}}{2a} = \frac{-2b}{2a} = -\frac{b}{a}.$$

8. The product of the roots of a quadratic equation is equal to c/a.

Proof: $$\left(\frac{-b + \sqrt{b^2 - 4ac}}{2a}\right)\left(\frac{-b - \sqrt{b^2 - 4ac}}{2a}\right)$$
$$= \frac{b^2 - (b^2 - 4ac)}{4a^2},$$

since the product in the numerators is that of a sum and difference and by Sec. 16 is equal to the square of the first part minus the square of the second. Furthermore,

$$\frac{b^2 - (b^2 - 4ac)}{4a^2} = \frac{b^2 - b^2 + 4ac}{4a^2} = \frac{4ac}{4a^2} = \frac{c}{a}.$$

9. It follows from (4) and (5) that if one of the roots of a quadratic equation is real, the other is real. It follows from (6) that if one of the roots is complex, then the other is also complex and is the conjugate of the first.

10. It follows from (7) and (8) above, and from the theorems I and II, Sec. 66, that a quadratic equation with conjugate complex roots will have real coefficients. Conversely, quadratic equations with real coefficients may have conjugate complex roots.

69. Applications of the Theory of a Quadratic Equation.

Example 1: Write the quadratic equation whose roots are $x = 2$ and $x = 3$.

Solution: If one thinks of the quadratic equation in the form $x^2 + \frac{b}{a}x + c/a = 0$, then the sum of the roots must equal $-b/a$; *i.e.*, $2 + 3 = 5 = -b/a$ so that $b/a = -5$. The product of the roots $= c/a = 2 \cdot 3 = 6$. Hence the equation is $x^2 - 5x + 6 = 0$.

Another method for finding this equation would be to change the equations giving the roots to factors, as follows. Since $x = 2$, then $x - 2 = 0$; and since $x = 3$, then $x - 3 = 0$. The product of the factors $(x - 2)(x - 3) = x^2 - 5x + 6 = 0$.

Example 2: Without solving, determine the nature of the roots of the equation $2x^2 - 5x + 7 = 0$.

Solution: The discriminant, $b^2 - 4ac$, in this case equals $(-5)^2 - 4 \cdot 2 \cdot 7 = 25 - 56 =$ a negative number. Hence the roots are complex. This also tells us that the graph of the equation would not cross the x-axis.

In problems involving the nature of the roots of a quadratic equation, one always computes the value of the discriminant, and draws conclusions based upon parts 3, 4, 5, and 6 of Sec. 68.

Example 3: Find the value of **k** in $2x^2 + 5x + k + 1 = 0$, if the difference of the roots is $\frac{7}{2}$.

Solution: Call the roots r_1 and r_2. According to (7), Sec. 68, $r_1 + r_2 = -\frac{5}{2}$. But $r_1 - r_2 = \frac{7}{2}$ as given. Adding these two relationships gives $2r_1 = 1$ or $r_1 = \frac{1}{2}$. Using this value of $r_1 = \frac{1}{2}$ we find that $\mathbf{r_2} = -3$. Thus we know the roots of the equation. From (8), Sec. 68, the product of the roots equals $\frac{k+1}{2}$ for the given equation. But $\frac{k+1}{2} = (\frac{1}{2})(-3) = -\frac{3}{2}$. This last relationship gives $k + 1 = -3$ or $k = -4$.

Many problems of this character are to be found, where a particular relationship involving the roots is described. The given relationship should be combined with either (7) or (8) of Sec. 68, thus providing sufficient data to enable one to solve the problem.

Example 4: Determine **k** so that the equation $3x^2 + kx - 2 = (x + 1)^2$ will have one root equal to -3.

Solution: Since -3 is a root, then -3 must satisfy the

given equation. That is, $3(-3)^2 - 3k - 2 = (-3 + 1)^2$ or $27 - 3k - 2 = 4$. This gives $k = 7$.
Using this value of **k**, the given equation can be reduced to $2x^2 + 5x - 3 = 0$, whose roots were shown in Example 3 to be $\frac{1}{2}$ and -3. Thus the conditions of Example 4 are satisfied.

70. Other Applications.

Certain problems require for their solution (or lead to) quadratic equations. These are discussed in Chapter XII.

These cases in general may be enumerated as follows:

a) Equations involving radicals,

b) Equations involving fractional indices,

c) Equations not themselves quadratic, but which are quadratic in form,

and the reader should consult Chapter XII for their discussion.

CHAPTER X

SYSTEMS OF QUADRATIC EQUATIONS

71. Introduction.

The general type form of a quadratic equation in two unknowns is usually written as: $ax^2 + bxy + cy^2 + dx + ey + f = 0$. The quantities a, b, ... f are constants, and one must make the restriction that not all three of the constants a, b, c are zero; otherwise the equation would be linear.

Two such equations may be solved simultaneously, as was the case for linear equations. However, a general method for solving such systems is not given in a first course in algebra. Instead, special methods for special cases are considered. We begin with the following intermediate case.

72. One Equation Linear, the Other Quadratic.

Illustration 1.

Solve simultaneously:

$$x^2 + xy - y^2 + 2x = -7 \qquad (1)$$
$$2x + 3y = 4. \qquad (2)$$

The following method is general. Solve the linear equation for one of the variables, say y, obtaining

$$y = \frac{4 - 2x}{3}. \qquad (3)$$

Substitute the value of y from (3) in (1), obtaining,

$$x^2 + x\left(\frac{4 - 2x}{3}\right) - \left(\frac{4 - 2x}{3}\right)^2 + 2x + 7 = 0.$$

This becomes upon simplification,

$$x^2 - 46x - 47 = 0 \text{ or } (x + 1)(x - 47) = 0.$$

These factors give solutions $x = -1$ or 47, and substituting them in (3) gives corresponding values of $y = 2$ or -30. These pairs of solutions are usually written, $(-1, 2)$ and $(47, -30)$.

Illustration 2.

Solve simultaneously
$$y = x^2 \qquad (1)$$
$$y = x + 2. \qquad (2)$$

Eliminating the value of y between these two equations gives

$$x^2 - x - 2 = 0. \qquad (3)$$

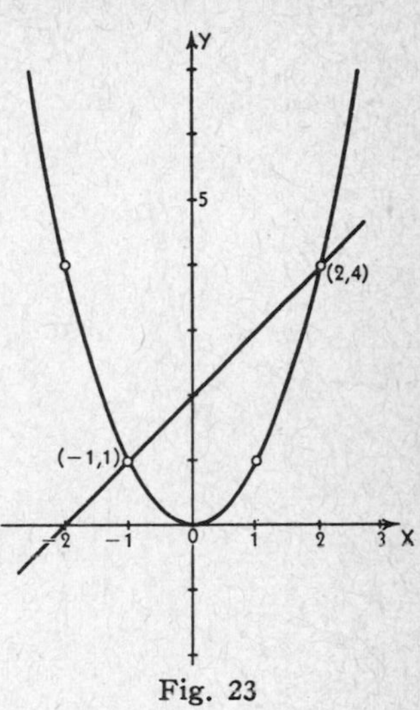

Fig. 23

Factoring, $(x - 2)(x + 1) = 0$. The solutions are $x = 2$ or -1, and the corresponding values of y are, $y = 4$ or 1. The graphs of the two equations are shown in Fig. 23. The simultaneous solutions represent the points of intersections of the two graphs. By examining the discriminant of equation (3) we see that $b^2 - 4ac = 1 + 8 = 9$, which tells us that the intersections will be distinct.

Illustration 3.

Solve simultaneously
$$y = x^2 \qquad (1)$$
$$y = x - \tfrac{1}{4}. \qquad (2)$$

Eliminating y from the two equations gives

$$x^2 - x + \tfrac{1}{4} = 0. \qquad (3)$$

Factoring $(x - \frac{1}{2})(x - \frac{1}{2}) = 0$. The two equal solutions are $x = \frac{1}{2}$ and $\frac{1}{2}$, giving $y = \frac{1}{4}$ and $\frac{1}{4}$. Fig. 24 shows that the parabola is tangent to the line at the point $(\frac{1}{2}, \frac{1}{4})$. This is to be expected since from equation (3), $b^2 - 4ac = 1 - 1 = 0$.

Fig. 24

Illustration 4.

Solve simultaneously
$$y = x^2 \qquad (1)$$
$$y = x - 3. \qquad (2)$$

Eliminating y gives the equation, $x^2 - x + 3 = 0$, (3)

whose roots are $x = \dfrac{1 + \sqrt{-11}}{2}, \dfrac{1 - \sqrt{-11}}{2}$,

and corresponding, $y = \dfrac{-5 + \sqrt{-11}}{2}, \dfrac{-5 - \sqrt{-11}}{2}$.

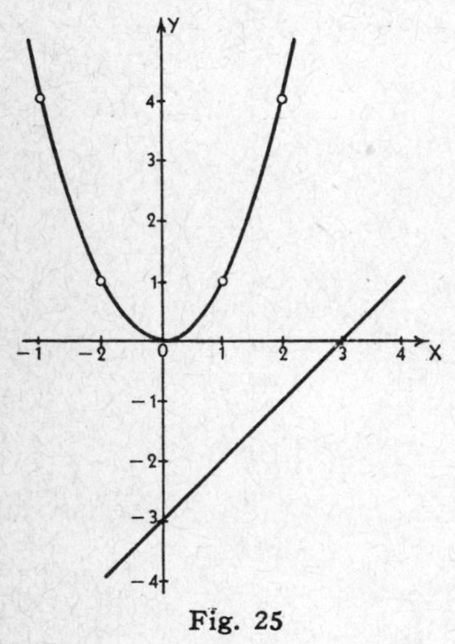

Fig. 25

Thus, the solutions are complex numbers, and the graphs **do not** intersect as shown in Fig. 25. This could have been anticipated, since for equation (3), $b^2 - 4ac = 1 - 12 = -11$.

73. Both Equations of Type $ax^2 + cy^2 + f = 0$.

Such a system may be solved by methods already used for linear systems. We illustrate by solving two problems.

Illustration 1.

Solve simultaneously

$$9x^2 + 16y^2 = 145 \quad (1)$$
$$3x^2 - 4y^2 = 11. \quad (2)$$

Using addition-subtraction method:
Multiply equation (2) by 4 and add to equation (1), obtaining

$$21x^2 = 189 \quad \text{or} \quad x^2 = 9.$$

Solving for x gives, $x = \pm 3$.
Now multiply equation (2) by 3 and subtract from equation (1) obtaining

$$28y^2 = 112 \quad \text{or} \quad y^2 = 4.$$

Solving for y gives, $y = \pm 2$. The solutions consist of all possible pairs of these values, namely, (3, 2), (3, −2), (−3, 2), (−3, −2). See Fig. 26.

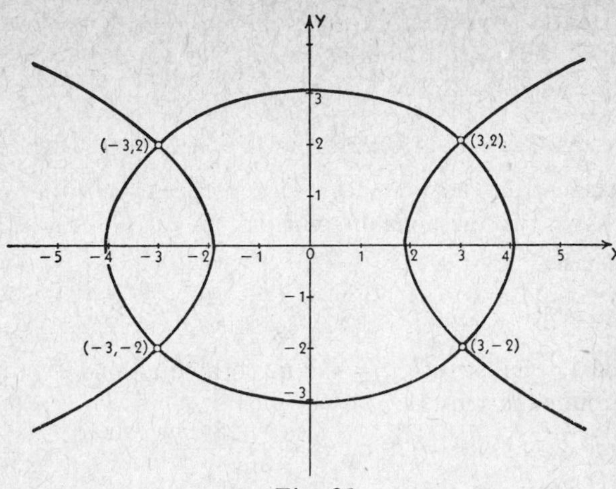

Fig. 26

Illustration 2.

Solve simultaneously $3x^2 - 4y^2 = 11$ (1)

$x^2 + y^2 = 13.$ (2)

Using substitution method:

Solving equation (2) for y^2, we have, $y^2 = 13 - x^2$. (3)

Substituting this value of y^2 in equation (1) gives

$$3x^2 - 52 + 4x^2 = 11 \quad \text{or} \quad x^2 = 9.$$

Solving for x gives, $x = \pm 3$. If either value of x is sub-

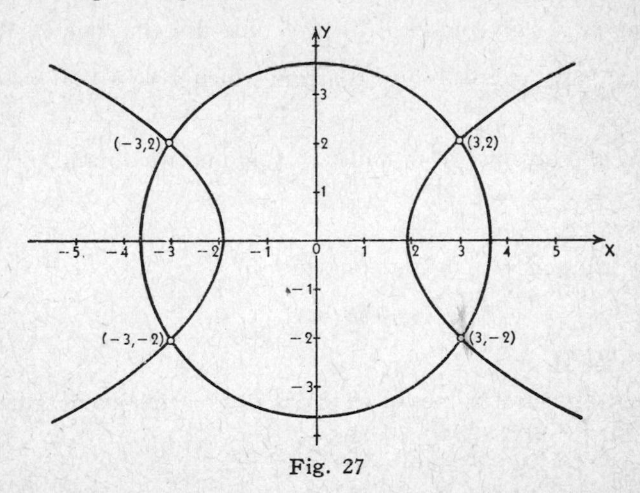

Fig. 27

stituted in equation (3), we obtain $y^2 = 4$ or $y = \pm 2$. The solutions are therefore, (3, 2), (3, −2), (−3, 2), (−3, −2). See Fig. 27.

74. All Terms Containing Unknowns Are of Second Degree.

This means that no first degree terms occur in either equation. We show two methods of solution.

Illustration 1.

Solve simultaneously $3xy + x^2 = 28$ (1)

$4y^2 + xy = 8.$ (2)

Method 1. Substitute $y = mx$ in both equations.

From equation (1):

$$3mx^2 + x^2 = 28 \text{ or } x^2 = \frac{28}{3m+1}. \quad (3)$$

From equation (2):

$$4m^2x^2 + mx^2 = 8 \text{ or } x^2 = \frac{8}{4m^2+m}. \quad (4)$$

Equating the two expressions for x^2 in (3) and (4), and clearing the equation of fractions,

$$112m^2 + 28m = 24m + 8 \text{ or}$$

$$28m^2 + m - 2 = 0. \quad (5)$$

Factoring equation (5): $(4m - 1)(7m + 2) = 0$, which yields

$$m = \tfrac{1}{4} \text{ and } m = -\tfrac{2}{7}. \quad (6)$$

Using $m = \frac{1}{4}$ in equation (3) [we could use equation (4)]

$$x^2 = \frac{28}{\frac{7}{4}} = 28 \cdot 4/7 = 16, \text{ from which } x = +4 \text{ or } -4.$$

Using these values of m and x in $y = mx$, we obtain

$y = \frac{1}{4} \cdot 4 = 1$ and

$y = \frac{1}{4} \cdot (-4) = -1.$

Now using $m = -2/7$ in equation (3)

$$x^2 = \frac{28}{1/7} = 28 \cdot 7 = 196, \text{ and}$$

$x = 14$ or -14.

Using this set of values of m and x in $y = mx$, we obtain

$y = -2/7(14) = -4$ and

$y = -2/7(-14) = +4.$

Value of m used	Corresponding values of x	$y = mx$
$m = \frac{1}{4}$	4 -4	1 -1
$m = -\frac{2}{7}$	14 -14	-4 4

The student will find it convenient to construct a chart as shown above to keep the values of m and the solutions correlated in the proper orders. The pairs of x and y cannot be combined indiscriminately, but must be paired as found in the stages of solution.

Method 2. Eliminating the independent constants.

Use the same equations, $3xy + x^2 = 28$ (1)

$4y^2 + xy = 8.$ (2)

Multiply equation (2) by 7: $28y^2 + 7xy = 56.$ (3)

Multiply equation (1) by 2: $6xy + 2x^2 = 56.$ (4)

Subtract (4) from (3): $28y^2 + xy - 2x^2 = 0.$ (5)

Factoring equation (5): $(4y - x)(7y + 2x) = 0.$ (6)

If each of the factors of (6) is solved with one of the equations (1) or (2), one obtains the solutions of the original system. Thus one may solve, by the method of Sec. 72, the two pairs of equations:

$$\begin{cases} 4y^2 + xy = 8 \\ 4y - x = 0 \end{cases} \quad \text{or} \quad \begin{cases} 4y^2 + xy = 8 \\ 7y + 2x = 0. \end{cases}$$

The reader should verify that these solutions are (4, 1), (−4, −1), (14, −4), (−14, 4), as were found by the first method of this section. Method 2 may be employed to advantage whenever the equation corresponding to (5) is readily factorable.

75. Symmetrical Equations.

If the interchanging of x and y leaves an equation unchanged, the equation is said to be **symmetrical** with respect to x and y. Thus, $3x^2 + xy + 3y^2 + 2x + 2y = 7$ is a symmetrical equation. It is easily seen that in order to have symmetry, x^2 and y^2 must have the same coefficients; also x and y must have the same coefficients.

Illustration 1.

Solve simultaneously $x^2 - xy + y^2 = 7$ (1)

$x + xy + y = 11.$ (2)

For this type, let $x = u + v$, $y = u - v$.

Substituting these values in (1) $u^2 + 3v^2 = 7.$ (3)

Substituting these values in (2) $u^2 + 2u - v^2 = 11.$ (4)

We can eliminate v^2 by adding 3 times (4) to (3), obtaining

$$4u^2 + 6u = 40.$$

which can be simplified to, $2u^2 + 3u - 20 = 0.$ (5)

Factoring (5), $(2u - 5)(u + 4) = 0$ so that $u = \frac{5}{2}$ or $u = -4$. Using $u = \frac{5}{2}$ in equation (3) we obtain $3v^2 = 7 - \frac{25}{4} = \frac{3}{4}$, from which $v^2 = \frac{1}{4}$, so that $v = \frac{1}{2}$ or $-\frac{1}{2}$. With these values of u and v, we obtain two pairs of values of x and y, as shown in the

Value of u	Corresponding values of v	$x = u + v$	$y = u - v$
$\frac{5}{2}$	$\frac{1}{2}$ $-\frac{1}{2}$	3 2	2 3
-4	$\sqrt{3}i$ $-\sqrt{3}i$	$-4 + \sqrt{3}i$ $-4 - \sqrt{3}i$	$-4 - \sqrt{3}i$ $-4 + \sqrt{3}i$

adjoining chart. Now using $u = -4$ in equation (3) we obtain $v^2 = -3$ so that $v = \pm\sqrt{3}i$. With this last set of values of u and v we obtain the remaining values of x and y. The substitution of $x = u + v$, $y = u - v$, may be used in solving many equations which are not symmetrical, but in which the signs + and − are the only aspects which interfere with symmetry.

76. Hints and Other Special Devices.

In this section the reader will find several typical cases solved. The illustrations which are given should not be looked upon as covering all of the varieties of such problems. They have been included with the view of helping the reader develop his ideas on the various tactics which may be employed to solve systems of equations. Most of the cases illustrated will lead to solutions involving only linear and quadratic equations. Somewhat similar devices may be used with equations of higher degree, but such cases are not studied in a first course in college algebra.

Illustration 1.

$x^2 + y^2 = 13$ (1)

$xy = 6.$ (2)

Multiply equation (2) by 2: $2xy = 12.$ (3)

Add (3) and (1): $x^2 + 2xy + y^2 = 25.$

Take square root of both sides: $x + y = \pm 5.$ (4)

Subtract (3) from (1): $x^2 - 2xy + y^2 = 1.$

Take square root: $x - y = \pm 1.$ (5)

Thus one is led to the problem of solving the equations (4) with (5). This means one has four systems of linear equations to solve, namely:

$$\begin{array}{llll} x + y = 5 & x + y = 5 & x + y = -5 & x + y = -5 \\ x - y = 1 & x - y = -1 & x - y = 1 & x - y = -1. \end{array}$$

The results in each set are respectively:

$$(3, 2) \qquad (2, 3) \qquad (-3, -2) \qquad (-2, -3).$$

The graphical interpretation of this method is as follows. The two equations, as given, intersect in four points. These four

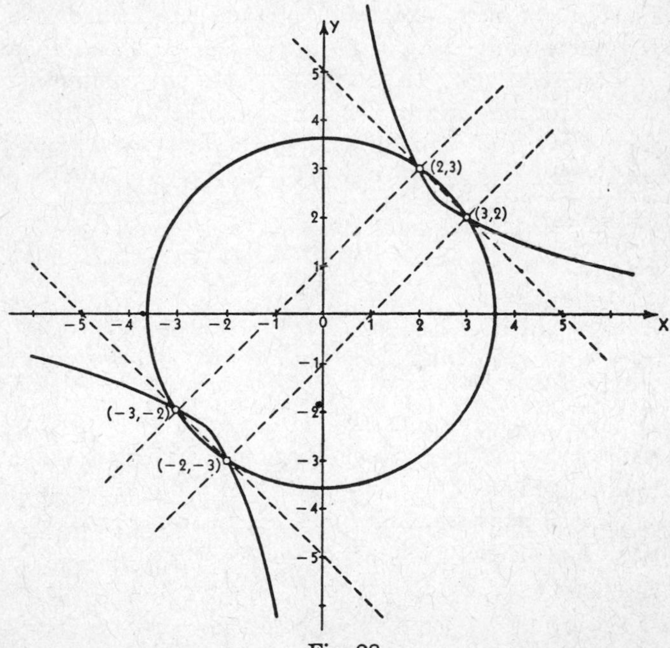

Fig. 28

points are also the positions in which the pairs of lines intersect. Fig. 28 shows the given curves in bold line, and the pairs of linear equations as dotted lines.

The points of intersection are common to the various systems.

Illustration 2.

$$x^3 - y^3 = 9 \qquad (1)$$

$$x - y = 3. \qquad (2)$$

Divide (1) by (2): $x^2 + xy + y^2 = 3.$ (3)

The real solutions of equations (2) and (3) are the same as the solutions of (1) and (2), and may be found by the method of Sec. 72.

Solving equation (2) for y gives $y = x - 3$.
Substituting this value for y in equation (3) gives:

$$x^2 + x^2 - 3x + x^2 - 6x + 9 = 3 \quad \text{or}$$

$$x^2 - 3x + 2 = 0.$$

The solutions for x are 2 or 1, with corresponding values for y being -1 or -2.

The points of intersection are therefore $(2, -1)$ and $(1, -2)$. Normally one would expect a third degree and a linear equation to have three solutions. In the present case one solution is infinite. In this outline we confine our attention to the real solutions, and leave the more general discussions for a course in algebraic geometry.

Illustration 3.

$$x^3 + y^3 = 35 \qquad (1)$$

$$x^2 - xy + y^2 = 7. \qquad (2)$$

Divide (1) by (2): $x + y = 5.$ (3)

Now solve (3) with (2) and one obtains two solutions, namely, (2, 3) and (3, 2).

The method used in Illustrations 2 and 3 may be used whenever the left hand member of one equation is the sum or difference of two cubes, and the left hand member of the other equation is a factor of the expression of the first equation. See Sec. 18, Formulas I and J, for the factors of these forms.

Illustration 4.

$$\sqrt[3]{x} + \sqrt[3]{y} = 5 \qquad (1)$$

$$x + y = 65. \qquad (2)$$

These two equations can be put in the same form as those of Illustration 2. For if one sets $\sqrt[3]{x} = u$ and $\sqrt[3]{y} = v$ in the above equations, they become respectively,

$$u + v = 5 \quad (3)$$

$$u^3 + v^3 = 65. \quad (4)$$

Now divide (4) by (3) $\quad u^2 - uv + v^2 = 13. \quad (5)$

Square equation (3) $\quad u^2 + 2uv + v^2 = 25. \quad (6)$

Subtract (5) from (6) $\quad 3uv = 12.$

Therefore, $\quad uv = 4$ or $v = 4/u. \quad (7)$

This value of v may now be substituted in equation (3), yielding,

$$u + 4/u = 5$$

$$\text{or } u^2 - 5u + 4 = 0$$

$$(u - 4)(u - 1) = 0.$$

Therefore, $\quad u = 4$ or $u = 1$.

From equation (7) the corresponding values of v are $v = 1$ or $v = 4$.

From the substitution which was made we see that $x = u^3$ and $y = v^3$. Consequently ($x = 64$, $y = 1$) and ($x = 1$, $y = 64$) are solutions of the given equations. The method just used to solve equations (3) and (4) is an alternative for solving Illustration 2 and the reader will find it valuable practice to solve the second illustration by this method.

Illustration 5.

$$x^3 - y^3 = 7xy \quad (1)$$

$$x - y = 2. \quad (2)$$

Divide equation (1) by equation (2):

$$x^2 + xy + y^2 = \tfrac{7}{2}xy. \quad (3)$$

Multiply (3) by 2 $\quad 2x^2 + 2xy + 2y^2 = 7xy. \quad (4)$

Collect terms $\quad 2x^2 - 5xy + 2y^2 = 0. \quad (5)$

From equation (2) $\quad x = y + 2. \quad (6)$

Substitute value of x from equation (6) in equation (5)

$$2(y^2 + 4y + 4) - 5y(y + 2) + 2y^2 = 0. \quad (7)$$

Collect terms $\quad y^2 + 2y - 8 = 0$

or $\quad (y + 4)(y - 2) = 0.$

Therefore $\quad y = 2$ or $y = -4$.

When these values of y are substituted in (6) $x = 4$ and $x = -2$. This gives the real solutions of the two equations, as (4, 2) and (−2, −4).

Illustration 6.

$$x + \sqrt{xy} + y = 14 \qquad (1)$$
$$x^2 + xy + y^2 = 84. \qquad (2)$$

Divide equation (2) by equation (1) obtaining

$$x - \sqrt{xy} + y = 6. \qquad (3)$$

Rewrite equation (1) $x + \sqrt{xy} + y = 14$

Adding $2x + 2y = 20$

or $x + y = 10. \qquad (4)$

Write equation (1) in the form $\sqrt{xy} = -(x + y) + 14. \qquad (5)$

Replace $(x + y)$ in (5) by value from (4)

$$\sqrt{xy} = -10 + 14 = 4$$

Squaring $xy = 16$

or $y = 16/x. \qquad (6)$

Substitute this value of y in (4)

$$x + 16/x = 10$$

or $x^2 - 10x + 16 = 0$

$$(x - 2)(x - 8) = 0.$$

Therefore $x = 2$ or 8.

From equation (6) $y = 8$ or 2

Therefore the solutions are (2, 8) and (8, 2).

Remark: The solutions may also be obtained by substituting the value of y from equation (4) into equation (2), giving: $x^2 + x(10 - x) + (10 - x)^2 = 84$ or $x^2 - 10x + 16 = 0$, as was also obtained above.

Illustration 7.

$$x^2 + y^2 = xy + 7 \qquad (1)$$
$$x - y = xy - 5. \qquad (2)$$

Add equation (1) and (2) $x^2 + y^2 + x - y = 2xy + 2 \qquad (3)$

Transpose and rearrange $x^2 - 2xy + y^2 + x - y - 2 = 0 \qquad (4)$

Rewrite (4) as $(x - y)^2 + (x - y) - 2 = 0 \qquad (5)$

Treat $(x - y)$ as a quantity and factor this quadratic equation into:

$$[(x - y) + 2] \cdot [(x - y) - 1] = 0,$$

yielding the two linear equations

$$x - y + 2 = 0 \qquad (6)$$

and $x - y - 1 = 0. \qquad (7)$

Now solve each of equations (6) and (7) in turn with equation (2).

From (6) and (2): $xy = 3$ or $y = 3/x$. (8)

Use the value of y in (6) $x - 3/x + 2 = 0$

or $x^2 + 2x - 3 = 0$

$(x + 3)(x - 1) = 0$

So that $x = -3$ or 1. From (8), $y = -1$ or 3.

Therefore two solutions are $(-3, -1)$ and $(1, 3)$.

Now from (7) and (2) $xy = 6$ or $y = 6/x$. (9)

Use this value of y in (7) $x - 6/x - 1 = 0$

or $x^2 - x - 6 = 0$

$(x - 3)(x + 2) = 0$

So that $x = 3$ or -2. From (9) $y = 2$ or -3.

Therefore two other solutions are $(3, 2)$ and $(-2, -3)$.

To summarize, the four solutions are $(-3, -1)$, $(1, 3)$, $(3, 2)$, $(-2, -3)$.

Illustration 8.

$$1/x + 1/y = 11 \quad (1)$$

$$1/x^2 + 1/y^2 = 61. \quad (2)$$

Two methods of solution will be shown.

Method 1. Square equation (1)

$$1/x^2 + 2/xy + 1/y^2 = 121$$

Repeat equation (2) $1/x^2 \quad + 1/y^2 = 61$

Subtract $2/xy \quad = 60.$ (3)

Now subtract (3) from (2) $1/x^2 - 2/xy + 1/y^2 = 1$

and since both sides of this equation are perfect squares, we have on extracting square roots,

$$1/x - 1/y = \pm 1. \quad (4)$$

Write (4) as two separate equations and solve each with (1). That is, solve each of the systems

$$\begin{cases} 1/x + 1/y = 11 \\ 1/x - 1/y = 1 \end{cases} \qquad \begin{cases} 1/x + 1/y = 11 \\ 1/x - 1/y = -1 \end{cases}$$

Adding	$2/x = 12$	$2/x = 10.$
Therefore	$x = 1/6$	$x = 1/5$
and	$y = 1/5$	$y = 1/6.$

Method 2. Let $1/x = u$, $1/y = v$ in the given equations and obtain:

$$u + v = 11$$
$$u^2 + v^2 = 61$$

These may be solved for u and v by the method of Sec. 72. From the values of u and v thus obtained, find the corresponding values of x and y.

Illustration 9.

$$x(x + y) = -6 \qquad (1)$$
$$y(x + y) = 10. \qquad (2)$$

Two methods will be shown for solving these equations.

Method 1. Divide (1) by (2),

$$x/y = -3/5 \text{ or } x = -\tfrac{3}{5}y. \qquad (3)$$

Substitute value of x from (3) in (1)

$$-3/5y(\tfrac{2}{5}y) = -6$$

or, dividing by -6, $\quad y^2/25 = 1$

or $\quad y^2 = 25$

so that $\quad y = +5 \text{ or } -5$

and from (3) $\quad x = -3 \text{ or } +3.$

The solutions are therefore $(-3, 5)$ and $(3, -5)$.

Method 2. Rewrite equations (1) and (2) as

$$x^2 + xy = -6 \qquad (1)$$
$$xy + y^2 = 10 \qquad (2)$$

Adding $\quad x^2 + 2xy + y^2 = 4$

so that $\quad x + y = \pm 2. \qquad (3)$

Equation (3), being equivalent to two equations, may be written

$$x + y = 2 \qquad (4)$$

and $\quad x + y = -2. \qquad (5)$

Replace $x + y$ in $x(x + y) = -6$ by its value 2 from equation (4), then

$$2x = -6$$
$$x = -3.$$

Therefore $\quad y = 5.$

Replacing $(x + y)$ by -2 from equation (5), one obtains

$$-2x = -6$$
$$x = 3$$

and $\quad y = -5.$

Illustration 10.

$$x^{1/2} + y^{1/2} = 5 \quad (1)$$

$$x^{-1/2} + y^{-1/2} = 5/6. \quad (2)$$

Let $x^{1/2} = \sqrt{x} = u$ and $y^{1/2} = \sqrt{y} = v$.

Then equations (1) and (2) become $\quad u + v = 5 \quad (3)$

$$1/u + 1/v = 5/6. \quad (4)$$

Reducing the left member of (4) to common denomination, we have,

$$\frac{u + v}{uv} = \frac{5}{6}. \quad (5)$$

But since $u + v = 5$ this becomes $\quad 5/uv = 5/6$

so that $\quad uv = 6. \quad (6)$

Now solve (3) with (6) obtaining

$$u(5 - u) = 6$$

$$u^2 - 5u + 6 = 0$$

$$(u - 3)(u - 2) = 0.$$

Therefore $\quad u = 3$ or 2

From which by either (3) or (6), $v = 2$ or 3.

Therefore $\quad \sqrt{x} = 3$ or 2, and $\sqrt{y} = 2$ or 3.

Upon solving for x and y we have the two pairs of solutions (9, 4) and (4, 9).

CHAPTER XI

THE NATURE OF PROBLEM SOLVING

77. Introduction.

In the previous chapters we have been giving special attention to definitions, the use of the fundamental operations, and the solution of linear and quadratic equations. Our work so far has been mainly the use of certain techniques.

Let us now consider some of the logical aspects involved in a study of algebra. In order to develop the ideas which we need for our present discussion, we must first consider some preliminary matters. This we do in the next three sections.

78. Notation and Symbols.

As the reader has already noticed, our work in algebra has been built around certain notations and symbols. We let numbers and quantities be represented by letters. Relationships connecting quantities are represented by equations; operations are represented by $+$, $-$, $\times$, $\div$, $\sqrt{\ }$, $\sqrt[3]{\ }$, a^n, etc.

Let us examine some special cases of the use of literal notation.

Suppose that **n** represents an integer. Then every **even** integer is represented by **2n,** as **n** takes on integral values.

There are several statements concerning this particular notation which we may summarize as follows.

If **n** is **any integer,** then

1) $2n$ always represents an **even** integer.
2) $2n + 1$, and $2n - 1$ are always **odd** integers.
3) $n, n + 1, n + 2, \ . \ . \ .$ etc., are **consecutive** integers. Another way of writing three consecutive integers would be $n - 1, n, n + 1$. Note that these two sequences are different for the same value of n. For, suppose that $n = 3$, then the first sequence has the values 3, 4, 5 but the second sequence has the values 2, 3, 4.

4) $2n$, $2n + 2$, $2n + 4$, or $2n$, $2(n + 1)$, $2(n + 2)$ are three **consecutive even** integers.
5) $2n + 1$, $2n + 3$, $2n + 5$, etc., are **consecutive odd** integers.

There are other examples which we might consider, but the above illustration is sufficient for our needs in this chapter.

79. From Words to Symbols.

The transfer from a written statement to an equation causes some students much trouble. Yet such a restatement is necessary in many problems in order to produce the equations which are to be solved. Practice in the writing of equations which represent the statement of the problem is practically the only means of overcoming such a difficulty.

Let us illustrate this transfer from words to symbols by means of several examples.

Illustration 1.

If x oranges cost 28 cents, express the cost of one orange. Since the total cost divided by the number of oranges represents the cost of a single orange, we have: One orange costs $28/x$ cents.

Illustration 2.

If a boy is of age x and his brother is three years older, express the sum of their ages.
If the age of the first boy is x, the age of his brother is $x + 3$. If S represents the sum of their ages,

$$S = x + x + 3 = 2x + 3.$$

Illustration 3.

Express the fact that Tom is three times as old now as he was ten years ago.
Let x = his present age;
then $x - 10$ = his age ten years ago.
By the statement of the problem $3(x - 10) = x$.
Note: For the solution we have $3x - 30 = x$ or $2x = 30$, $x = 15$, and this checks with the statement as given.

Illustration 4.

Find the distance between two air fields if a plane travelling x miles per hour requires h hours to make the journey.

The use of literal values as above frequently confuses students. The same sort of problem, if stated with numerical values, usually presents no difficulty. Thus, if one is told that the plane travels at a rate of 120 m.p.h. for 4 hours, the distance is readily seen to be $4(120) = 480$ miles. By analogy, the required distance in the problem is $h \cdot x$ miles.

80. Reasoning.

There are two principal types of reasoning usually applied in solving problems algebraically. These are known as deductive and inductive reasoning. We shall confine our present discussion to a simple study of the deductive process. Inductive reasoning is discussed in Chapter XVI.

We define **deductive** reasoning by saying that in this type one starts with some premise* (or assumption), and by logical steps deduces from the given assumption a result which is called the conclusion. In algebra, the premise is often obtained from the statement of the given problem. The argument is based upon the axioms of equality (see Sec. 14) and the fundamental operations, etc. Perhaps the whole matter will be clearer after a few illustrations.

Illustration 1.

Find a value of x such that $x - 3 = \sqrt{x + 3}$.

Assumption: The given equation expresses the assumption that a value of x exists for which the equation will be true.

Argument: Starting with the given equation, we apply Axiom 6, Sec. 14, and square both sides of the equation, thus obtaining,

$$x^2 - 6x + 9 = x + 3.$$

Applying Axiom 2, Sec. 14, we subtract $x + 3$ from both sides of the equation, obtaining

$$x^2 - 7x + 6 = 0.$$

* In many problems which lead to the solution of a simultaneous system of equations there will be more than one assumption expressed by the equations of condition.

Factoring this quadratic expression according to the principles of Sec. 19, we have

$$(x - 6)(x - 1) = 0.$$

Since the product of the two factors is equal to zero, one of the factors must have the value zero. If the first factor is zero then x must equal 6. If the second factor is zero then x must equal 1.

Conclusion: We have arrived at the conclusion that x must equal 6 or 1. These values must now be tested in the original equation. If $x = 6$, then $6 - 3 = \sqrt{6 + 3}$ or $3 = 3$. If $x = 1$, then $1 - 3 = \sqrt{1 + 3}$ or -2 must equal $\sqrt{4}$ or 2. But -2 does not equal $+2$. Hence, we conclude that the only value which satisfies the given equation is $x = 6$.

Illustration 2.

A man A in an automobile is traveling at the rate of 55 miles per hour. A state trooper B starts out one hour later to overtake him in 4 hours. How fast must B travel to overtake A?

Assumption: Let x = miles per hour at which B travels.

Argument: $5(55) = 275$ miles, or the distance A has traveled when B overtakes him. But by assumption $4x =$ the number of miles B will travel. At the time of overtaking these two distances will be equal; hence

$$4x = 275.$$

Divide both sides of this equation by 4 and according to Axiom 4, Sec. 14,

$$x = 68\tfrac{3}{4}.$$

Conclusion: The rate at which B must travel is $68\frac{3}{4}$ miles per hour.

After considering these two illustrations we may draw some general conclusions. The first illustration led us to two numerical results, one of which satisfied the given equation, the other of which did not. In the second illustration only one solution was obtained, and this satisfied the given statement of the problem.

The situation which arose in the first illustration is explained in the following way. We did not find our solutions directly from the given equation. Instead, we performed certain operations which finally resulted in the equation $x^2 - 7x + 6 = 0$. From this we obtained two solutions, $x = 6$ and $x = 1$. Both of these solutions satisfy the equation $x^2 - 7x + 6 = 0$, but only one of them, namely $x = 6$, satisfies the given equation. Consequently we may rightly ask, under what conditions will the solutions of the derived equation satisfy the given equation? The answer to this question will be found in the next few sections.

81. Equivalent Equations.

Two equations are **equivalent** if every solution of either of them is a solution of the other also. In the preceding section we obtained a derived equation from a given equation by performing a sequence of algebraic operations. The following operations lead to **equivalent** derived equations.

a) Adding or subtracting the same quantity to both sides of the equation.

b) Multiplying or dividing both sides of an equation by the same constant numerical value, provided the constant is different from zero.

82. Redundant and Defective Equations.

If the derived equation has more solutions than the given equation, the derived equation is said to be **redundant.** Thus the derived equation in Illustration 1, Sec. 80, is redundant. A value such as $x = 1$ in that illustration is called **extraneous.**

The following operations may lead to derived equations which are **redundant.**

a) Multiplying both sides of an equation by a factor which contains the unknown quantity.

b) Raising both sides of the given equation to the same power.

If the derived equation contains fewer solutions than the given equation, the derived equation is said to be **defective.**

The following operations may lead to derived equations which are **defective.**

c) Dividing both sides of an equation by a factor which contains the unknown quantity.

d) Extracting the same root of both sides of an equation.

Illustration 1.

Solve the equation $\sqrt{x+4} + 1 = \sqrt{x-1}$.

Solution:

Squaring both sides: $x + 4 + 2\sqrt{x+4} + 1 = x - 1$
Collecting terms: $2\sqrt{x+4} = -6$
Dividing by 2: $\sqrt{x+4} = -3$.
Squaring both sides: $x + 4 = 9$
Solving for x: $x = 5$.

Checking: If $x = 5$ is substituted in the original equation, we have: $\sqrt{9} + 1 = \sqrt{5-1}$ or $3 + 1 = 2$ which is obviously false.

Conclusion: The result obtained is extraneous. It satisfies the derived equation but not the original equation. Since the only solution obtained does **not** satisfy the **given equation** we must conclude that there are no values of x for which the given equation is true.

This does not mean that there is some defect in our algebraic solution. It means that we had an assumption, namely the given equation. By deduction, we arrived at the conclusion that if there is a value of x satisfying the given assumption it might be $x = 5$. But $x = 5$ does not satisfy the given relationship. Hence the assumption was false, and we have been able to detect this fact by our deductive process.

Illustration 2.

Solve the equation $x^2 - 5x + 6 = 2x - 4$.

Solution: Rewrite the given equation in the form

$$(x - 3)(x - 2) = 2(x - 2).$$

Divide both sides by $x - 2$ obtaining

$$x - 3 = 2 \text{ from which}$$
$$x = 5.$$

Comments: The derived equation is defective since it has only one solution whereas the original equation has two solutions, namely $x = 5$ and $x = 2$. It is possible to solve the given equation without losing any solutions. This is done as follows. Write the equation as:

$$(x - 3)(x - 2) - 2(x - 2) = 0.$$

Then factor into:

$$(x - 2)[(x - 3) - 2] = (x - 2)(x - 5) = 0.$$

These two factors contribute the two solutions $x = 5$ and $x = 2$, because the original equation was factored instead of being divided by $x - 2$ and losing one of the solutions.

83. Checking Solutions.

We have discovered in this chapter the reasons for and the necessity for checking all solutions to algebraic problems. The student should form the habit of checking every solution to a problem in order that extraneous values may be discarded and that lost solutions may be recovered.

Note: When checking solutions to problems, always substitute the values of the unknown in the original equation.

CHAPTER XII

SOLUTIONS OF TYPICAL PROBLEMS

84. Introduction.

There are many types of problems which may be solved by algebra. It would be almost impossible to show the solutions of every type, and for that reason only a limited number of cases are considered in this chapter. They have been chosen because: a) they represent types which often give students a considerable amount of trouble; b) they represent certain applications of the theory; and c) they should give the student some idea of the methods of setting up equations from the stated problems.

85. Equations Involving Fractional Indices.

Illustration 1.

Solve $x^{2/3} + x^{1/3} - 6 = 0$.

Solution: Let $x^{1/3} = y$; then the equation becomes

$$y^2 + y - 6 = 0.$$

Factoring, $(y - 2)(y + 3) = 0$.

Solving, $y = 2$ or $y = -3$.

Therefore, $x^{1/3} = 2$ or $x^{1/3} = -3$.

Cubing, $x = 8$ or $x = -27$.

Substituting these values of x in the given equation, we find that they both check and are therefore solutions.

Illustration 2.

Solve for x, if $x^{3/2} = 8$.

Solution: Extract the cube root of both sides of the equation, obtaining $x^{1/2} = 2$. Now square both sides, and $x = 4$. [Read Sec. 33, on Fractional Exponents.]

86. Equations Quadratic in Form.

Equations such as $x^4 + 5x^2 + 6 = 0$, $x^{3/2} - 13x^{3/4} + 36 = 0$, or in general, $ax^{2n} + bx^n + c = 0$, where n is any integer or fraction, are said to be quadratic in form. The following examples will show how such equations may be solved by the methods used in Chapter IX for solving simple quadratic equations.

Illustration 1.

Solve $x^6 - 7x^3 - 8 = 0$.

Solution: Let $x^3 = y$, and the equation becomes

$$y^2 - 7y - 8 = 0.$$

Factoring, $(y - 8)(y + 1) = 0$.

Hence, $y = 8$ or $y = -1$.

Therefore, $x^3 = 8$ or $x^3 = -1$.

These last two equations may be written as,

$$x^3 - 8 = 0 \text{ and } x^3 + 1 = 0,$$

and are solved by factoring according to Formulas I and J of Sec. 18.

From $x^3 - 8 = 0$ we have, upon factoring,

$$(x - 2)(x^2 + 2x + 4) = 0$$

and the first factor gives $x = 2$.

The second factor yields by the quadratic formula,

$$x = -1 \pm \sqrt{-3}.$$

From $x^3 + 1 = 0$ we have by factoring,

$$(x + 1)(x^2 - x + 1) = 0.$$

The first factor gives $x = -1$. The second factor gives

$$x = \frac{1 \pm \sqrt{-3}}{2}.$$

There are therefore six solutions to the given equation, namely, $x = +2$, $x = -1$, $x = -1 + \sqrt{-3}$, $x = -1 - \sqrt{-3}$, $x = \frac{1 + \sqrt{-3}}{2}$ and $x = \frac{1 - \sqrt{-3}}{2}$.

Illustration 2.

Solve $x^{-1} - 4x^{-1/2} + 3 = 0$.

Solution: Let $x^{-1/2} = y$. Then we may write,

$$y^2 - 4y + 3 = 0.$$

Factoring: $(y - 1)(y - 3) = 0$.

Therefore, $y = 1$ or $y = 3$

So that: $x^{-1/2} = 1$ or $x^{-1/2} = 3$
and, $x^{1/2} = 1$ or $x^{1/2} = 1/3.$
Therefore $x = 1$ or $x = 1/9.$
Both of these values satisfy the given equation.

Illustration 3.

Solve $\sqrt{\dfrac{x^2+3}{x}} - \sqrt{\dfrac{x}{x^2+3}} = 3/2.$

Solution: Let $\sqrt{\dfrac{x^2+3}{x}} = y.$ Then

$$y - 1/y = 3/2.$$

Multiplying by 2y: $2y^2 - 3y - 2 = 0$
Factoring: $(2y+1)(y-2) = 0$
$y = -\tfrac{1}{2}$ or $y = 2$

From the first value of y, $\sqrt{\dfrac{x^2+3}{x}} = -\tfrac{1}{2}$

Squaring: $\dfrac{x^2+3}{x} = \tfrac{1}{4}$

Clearing fractions and transposing, we have:

$$4x^2 - x + 12 = 0$$

By quadratic formula, $x = \dfrac{1 \pm \sqrt{-191}}{8}$
and both of these solutions prove to be extraneous.

From y = 2, we have $\sqrt{\dfrac{x^2+3}{x}} = 2.$

Squaring, $\dfrac{x^2+3}{x} = 4$

Clearing fractions and transposing, we have

$$x^2 - 4x + 3 = 0$$

Factoring: $(x-3)(x-1) = 0$
Giving as solutions, $x = 3$ or $x = 1.$
Both of these values **satisfy** the given equation.

Illustration 4.

Solve $x^2 + 1/x^2 + x + 1/x = 4.$

Solution: This equation in its present form is not exactly of the type we are discussing. However, if we add 2 to both sides we may rewrite it as,

$$(x^2 + 2 + 1/x^2) + (x + 1/x) = 6$$

or $(x + 1/x)^2 + (x + 1/x) - 6 = 0.$

Now let $(x + 1/x) = y$. Then,

$$y^2 + y - 6 = 0$$
$$(y + 3)(y - 2) = 0.$$

Therefore, $y = -3$ or $y = 2$.

These values of y yield respectively,

$x + 1/x = -3$	$x + 1/x = 2$
$x^2 + 3x + 1 = 0$	$x^2 - 2x + 1 = 0$
$x = \dfrac{-3 \pm \sqrt{5}}{2}$	$(x - 1)(x - 1) = 0$
	$x = 1, \quad x = 1.$

The four solutions are, $x = 1$, $x = 1$, $x = \dfrac{-3 + \sqrt{5}}{2}$,

$$x = \frac{-3 - \sqrt{5}}{2}.$$

Illustration 5.

Solve $3x^2 - 4x + \sqrt{3x^2 - 4x - 6} = 18.$

Solution: If 6 is subtracted from each side of this equation, one obtains

$$(3x^2 - 4x - 6) + \sqrt{3x^2 - 4x - 6} = 12,$$

which is now in quadratic form.

Let $\sqrt{3x^2 - 4x - 6} = y$; then

$$y^2 + y - 12 = 0$$
$$(y - 3)(y + 4) = 0$$

and $y = 3$ or $y = -4$.

Therefore,

$\sqrt{3x^2 - 4x - 6} = 3$	or $\sqrt{3x^2 - 4x - 6} = -4$
$3x^2 - 4x - 6 = 9$	$3x^2 - 4x - 6 = 16$
$3x^2 - 4x - 15 = 0$	$3x^2 - 4x - 22 = 0$
$(3x + 5)(x - 3) = 0$	$x = \dfrac{4 \pm \sqrt{280}}{6} = \dfrac{4 \pm 2\sqrt{70}}{6}$
$x = -5/3$ or $x = 3$	$x = \dfrac{2 \pm \sqrt{70}}{3}.$

The values $x = -5/3$ and $x = 3$ satisfy the given equation but the other two values of x are extraneous.

87. Equations Involving Fractions.

The general attack on such problems is to multiply both sides of the equation by the L.C.M. of the denominators of the fractions.

Illustration 1.

Solve: $$\frac{5x - 3}{2x - 1} = \frac{3x + 1}{x + 1}.$$

Solution: Multiply both sides of the equation by

$$(2x - 1)(x + 1);$$

then, $(5x - 3)(x + 1) = (3x + 1)(2x - 1)$.
Perform the multiplication, thus obtaining,

$$5x^2 + 2x - 3 = 6x^2 - x - 1.$$

Collecting terms, $x^2 - 3x + 2 = 0$

$$(x - 1)(x - 2) = 0$$

giving $x = 1$ or $x = 2$.

Both of these values satisfy the given equation.

Illustration 2.

Solve $$\frac{1}{x - 2} = \frac{2x - 4}{x^2 - 5x + 6}.$$

Solution: Multiply both sides by $x^2 - 5x + 6$;
then $x - 3 = 2x - 4$.
Collecting terms, $x = +1$.

Remark: If one should clear fractions by forming the product of means and extremes, one obtains

$$x^2 - 5x + 6 = 2x^2 - 8x + 8.$$

Collecting terms, $x^2 - 3x + 2 = 0$

$$(x - 1)(x - 2) = 0.$$

Therefore $x = +1$ or $x = 2$.

The value of $x = +1$ checks as before. The value $x = 2$ is extraneous, for it causes the denominator of each fraction to be zero. The student should read Secs. 6 and 82 in connection with this last remark.

Illustration 3.

Solve $\frac{2x}{2x-1}+\frac{4}{8x-4}=\frac{1}{2}$.

Solution: Frequently it saves labor to reduce all fractions to lowest terms before solving the equation. Thus the second fraction in the above equation may be written

$$\frac{1}{2x-1}.$$

The equation then becomes

$$\frac{2x}{2x-1}+\frac{1}{2x-1}=\frac{1}{2}.$$

Clearing fractions, $4x+2=2x-1$

$$2x=-3$$
$$x=-3/2.$$

88. Equations Involving Radicals.

The general procedure for these problems is to isolate one radical at a time to one side of the equation. Then upon squaring both sides of the equation, all terms arising from the isolated radical are free from the radical. This process is repeated until all radicals are eliminated, and the resulting equation is solved by the usual methods.

Extraneous solutions may arise, and therefore each solution obtained must be checked in the original equation. The following examples should make the procedure clear.

Illustration 1.

Solve $x+2\sqrt{x+3}=21$.

Solution: Transposing, $2\sqrt{x+3}=21-x$

Squaring, $4(x+3)=441-42x+x^2$,

Collecting terms, $x^2-46x+429=0$,

Factoring, $(x-13)(x-33)=0$.

Therefore, $x=13$ or $x=33$.

The value $x=13$ satisfies the given equation but $x=+33$ does not and is extraneous.

Illustration 2.

Solve $\sqrt{13 + 4\sqrt{x - 1}} = 5$.

Solution: Square both sides; then

$$13 + 4\sqrt{x - 1} = 25$$

or $$4\sqrt{x - 1} = 12,$$

or $$\sqrt{x - 1} = 3.$$

Squaring again, $$x - 1 = 9$$

or $x = 10$, and this value satisfies the equation.

Illustration 3.

Solve $\sqrt{x} + \sqrt{x + 1} - \sqrt{2x + 1} = 0$.

Solution: Transpose the third radical to the right side of the equation, obtaining

$$\sqrt{x} + \sqrt{x + 1} = \sqrt{2x + 1}. \quad (1)$$

Squaring both sides, we obtain

$$x + 2\sqrt{x}\sqrt{x + 1} + x + 1 = 2x + 1.$$

(Note that the square of the left hand member of equation (1) involves the square of a binomial quantity. According to Sec. 16, Formula A, upon squaring we should have the square of the first radical plus twice the product of the radicals, plus the square of the second radical.)

Collecting terms, $2\sqrt{x(x + 1)} = 0$ or

$$\sqrt{x(x + 1)} = 0.$$

Squaring again, $$x(x + 1) = 0.$$

Therefore, $x = 0$ or $x = -1$.

Check: for $x = 0$, $0 + \sqrt{1} - \sqrt{1} = 0$

for $x = -1$, $\sqrt{-1} + 0 - \sqrt{-1} = 0$.

Although the result for $x = -1$ leads to imaginary numbers, we have a case in which both solutions satisfy the given equation.

Illustration 4.

Solve $\sqrt{x + 10} - \dfrac{6}{\sqrt{x + 10}} = 5$.

Solution: This equation may be solved either by the method of Sec. 86, or as follows:

Clearing of fractions, $x + 10 - 6 = 5\sqrt{x + 10}$, or

$$x + 4 = 5\sqrt{x + 10}.$$

Squaring both sides, $x^2 + 8x + 16 = 25(x + 10)$

Collecting terms, $x^2 - 17x - 234 = 0$

Factoring, $(x - 26)(x + 9) = 0$

Therefore, $x = 26$ or $x = -9$

The value $x = 26$ checks, but $x = -9$ is extraneous.

89. Uniform Rate Problems.

This class of problems depends upon the relation between distance, rate, and time.

If **d** represents distance, **r** represents rate, and **t** represents time, then these three quantities are related as follows:

$$\mathbf{d = r \cdot t.}$$

This relationship may be solved for **r** or **t**, thus giving

$$\mathbf{r = d/t} \quad \text{and} \quad \mathbf{t = d/r}.$$

Illustration 1.

A man can row downstream 10 miles and return in 6 hours. The rate of the stream is 4 miles per hour. Find his rate of rowing in still water.

Solution: Let x = rate the man rows in still water.

Then $x + 4$ = rate of travel downstream

and $x - 4$ = rate of travel upstream.

$$\frac{10}{x + 4} = \text{time it takes going downstream.}$$

$$\frac{10}{x - 4} = \text{time it takes going upstream.}$$

Since the sum of these times equals 6 hours,

$$\frac{10}{x + 4} + \frac{10}{x - 4} = 6$$

$$10(x - 4) + 10(x + 4) = 6(x^2 - 16)$$

$$10x - 40 + 10x + 40 = 6x^2 - 96$$

$$3x^2 - 10x - 48 = 0$$

$$(x - 6)(3x + 8) = 0$$

$$x = 6 \text{ or } -\tfrac{8}{3}.$$

The physical nature of this problem prohibits the use of $-\frac{8}{3}$. Hence his rate of rowing is 6 miles per hour.

Illustration 2.

A car travelled a certain distance at a uniform rate. If the speed of the car had been 10 miles per hour more, the journey would have occupied 3 hours less. Had the speed been 5 miles per hour less, the journey would have occupied 2 hours more. Find the distance and the speed.

Solution: Let x = the speed of the car in miles per hour.
y = the time of the journey.
Then xy = the distance travelled.

By the first condition the speed is $x + 10$, the time $y - 3$. By the second condition the speed is $x - 5$, the time $y + 2$.

The equations may be set up for the distance as expressed by each condition.

1st case: (distance) $xy = (x + 10)(y - 3)$ (1)

2nd case: $xy = (x - 5)(y + 2)$. (2)

From equation (1) we have,

$$xy = xy - 3x + 10y - 30 \quad \text{or}$$
$$-3x + 10y = 30. \qquad (3)$$

From equation (2) we have,

$$xy = xy + 2x - 5y - 10 \quad \text{or}$$
$$2x - 5y = 10. \qquad (4)$$

Solving simultaneously equations (3) and (4) by any method from Chapter VIII, we find

$$x = 50 \quad \text{and} \quad y = 18.$$

Hence the speed is 50 miles per hour, the time is 18 hours, and the distance 900 miles.

90. Mixture Problem.

Illustration 1.

The radiator of a car contains 22 quarts of a mixture of water and alcohol, the alcohol comprising 25%. How much of the mixture must be drawn off and replaced by alcohol so that the radiator will contain a mixture of which 50% is alcohol?

Solution: Let x = number of quarts of alcohol to be added.
Then $22 - x$ = number of quarts of liquid remaining after drawing off x quarts of the mixture.
Use the fraction $\frac{1}{4}$ to represent 25%.
Then, $\frac{22 - x}{4}$ = amount of alcohol in the remaining mixture.
$\frac{22 - x}{4} + x$ = amount of alcohol in the new mixture.
But 50% of 22 quarts, or 11 quarts, = the amount of alcohol in the new mixture.

Hence,
$$\frac{22 - x}{4} + x = 11$$
$$22 - x + 4x = 44$$
$$3x = 22$$
$$x = \tfrac{22}{3} \text{ or } 7\tfrac{1}{3} \text{ quarts.}$$

Check: After $7\frac{1}{3}$ quarts are drawn from 22 quarts there remain $14\frac{2}{3}$, or $\frac{44}{3}$ quarts, of which $\frac{1}{4}$, or $\frac{44}{12}$ quarts, are alcohol. To this is added $\frac{22}{3}$ quarts of alcohol so that the new mixture contains
$\frac{44}{12} + \frac{22}{3} = \frac{44}{12} + \frac{88}{12} = \frac{132}{12} = 11$ quarts of alcohol. Thus the solution is checked.

91. Work Problem.

Illustration 1.

It takes B twice as long as it takes A to do a certain piece of work. Working together, they can do the work in 6 days. How long would it take A to do it alone?

Solution: Let x = the number of days it takes for A to do the work.
Then 2x = the number of days it takes for B to do the work.
$1/x$ = the fractional part of the work done by A in 1 day.

$1/2x$ = the fractional part of the work done by B in 1 day.

$1/x + 1/2x$ = the fractional part of the work done by both in 1 day.

By the statement of the problem, working together they can do $\frac{1}{6}$ of the work in one day; hence,

$$1/x + 1/2x = 1/6.$$

Clearing fractions, $6 + 3 = x$ or

$$x = 9.$$

Hence A can do the work alone in 9 days.

Check: A does $\frac{1}{9}$, and B does $\frac{1}{18}$ of the work in one day. Together they do $\frac{1}{9} + \frac{1}{18} = \frac{3}{18} = \frac{1}{6}$ of the work in one day. Hence it will take them 6 days to complete the work.

92. Age Problem.

Illustration 1.

A man is now three times as old as his son. In ten years he will be twice as old as his son. Find their present ages.

Solution: Let x = the present age of the son.

Then $3x$ = the present age of the father.

$x + 10$ = the son's age ten years hence.

$3x + 10$ = the father's age ten years hence.

From the statement of the problem,

$$3x + 10 = 2(x + 10)$$

$$3x + 10 = 2x + 20$$

$$x = 10$$

$$\text{and } 3x = 30.$$

Consequently, the son is 10 years old and the father is 30 years old.

Note: A solution may also be obtained if one let

x = present age of the father,

$x/3$ = present age of the son.

To do so produces an equation involving fractional coefficients. This is of no particular disadvantage except that an extra step is involved in solving the equations when one rids the equation of fractions.

93. Problems Concerning Levers.

Problems concerning the lever are based upon the following formula, which expresses the physical law of the lever:

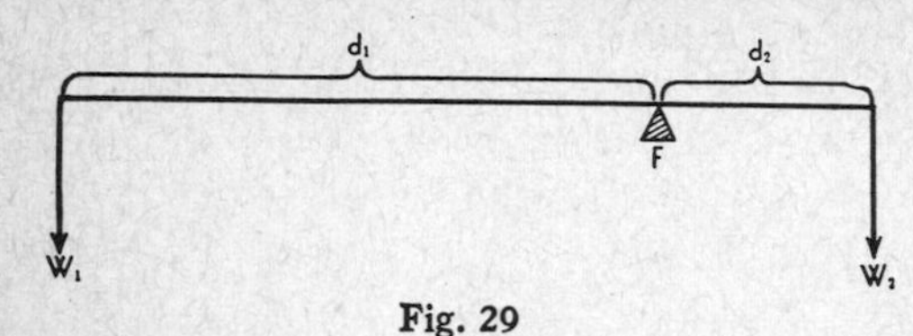

Fig. 29

$$W_1d_1 = W_2d_2,$$

where d_1 and d_2 are the lengths of the lever arms from the point of support (or fulcrum) and W_1 and W_2 are the weights (or forces) which are applied.

Illustration.

A man weighs 180 pounds. He wishes to lift a weight of 1200 lbs. by means of a lever. How far from the object must he place the fulcrum if one disregards the weight of the lever, which is six feet long?

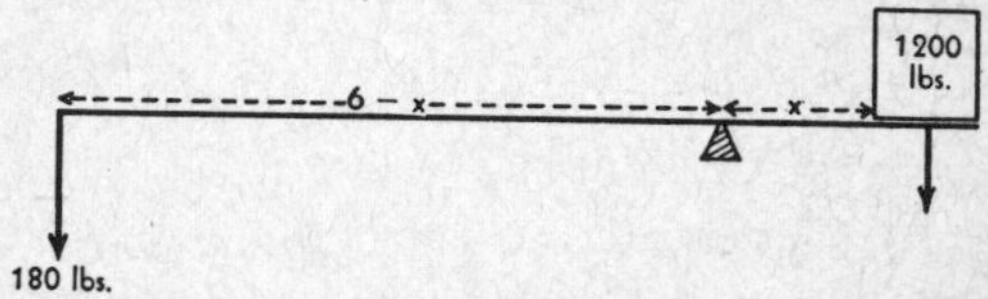

Fig. 30

Solution:

Using the law from the principle of the lever as explained above, we have the equation,

$$(6 - x)180 = 1200x$$

$$1080 - 180x = 1200x$$

$$1080 = 1380x$$

$$x = \frac{1080}{1380} = \text{approx. } .78 \text{ ft.} = 9.36 \text{ inches.}$$

Hence the fulcrum must be placed about .78 ft., or 9.36 inches, from the 1200 lb. weight which is to be lifted.

94. Clock Problems.

Illustration 1.

In how many minutes after 8 o'clock will the minute hand overtake the hour hand?

Solution: Let x = the number of minutes after 8 o'clock in which the minute hand overtakes the hour hand.

Since the hour hand moves only $\frac{1}{12}$ as fast as the minute hand, it will move x/12 minutes while the minute hand overtakes it. Therefore, in x minutes the minute hand must move over 40 more divisions than the hour hand. This is expressed in equation form as,

$$x - x/12 = 40$$
$$12x - x = 480$$
$$11x = 480.$$

Therefore, $x = 43\frac{7}{11}$ minutes.

Illustration 2.

At what times between 4 and 5 o'clock will the hands of a clock be at right angles?

Note: There will be two solutions to the problem:

a) When the minute hand lacks 15 minutes of overtaking the hour hand,

b) When the minute hand is 15 minutes past the hour hand.

Solution:

Case (a).

Let x = the number of minutes past 4 o'clock when the minute hand is 15 minutes behind the hour hand.

Then x/12 will represent the movement of the hour hand in these x minutes. Now, since the hour hand has a start of 20 minutes on the minute hand, we may express their relative positions as,

$$x + 15 = x/12 + 20 \quad \text{or}$$
$$x = x/12 + 5$$
$$12x = x + 60$$
$$11x = 60$$
$$x = 5\tfrac{5}{11} \text{ minutes past 4 o'clock.}$$

Case (b).

Let x = the number of minutes past 4 o'clock when the minute hand is 15 minutes ahead of the hour hand.

Their relative positions are then expressed by,

$$x = x/12 + 20 + 15$$
$$x = x/12 + 35$$
$$12x = x + 420$$
$$11x = 420$$
$$x = 38\tfrac{2}{11} \text{ minutes past 4 o'clock.}$$

95. Digit Problem.

Illustration 1.

A certain number consists of three digits. If the first and third digits are interchanged, the number is increased by 396. The sum of the digits is 9. Three times the hundred's digit is equal to the ten's digit. Find the number.

Remark: The notation for these problems is based upon the fact that a number such as 327 means $3(100) + 2(10) + 7$.

Solution: Let x = the hundred's digit
y = the ten's digit
z = the unit's digit.

Then $100x + 10y + z$ represents the number.
If the first and third digits are interchanged, the new number will be represented by $100z + 10y + x$. By the first condition of the problem,

$$100z + 10y + x = 100x + 10y + z + 396. \qquad (1)$$

By the second condition,

$$x + y + z = 9. \qquad (2)$$

By the third condition,

$$3x = y. \qquad (3)$$

These three equations may be simplified to

$$x - z = -4$$
$$x + y + z = 9$$
$$3x - y = 0, \text{ respectively.}$$

Solving this system of equations by the method of Sec. 62, we find that $x = 1$, $y = 3$, $z = 5$. Therefore, the number is 135.

CHAPTER XIII

INEQUALITIES

96. Definition and Symbols.

An **inequality** is a statement of the fact that one quantity is greater than or less than another quantity. The idea of inequality applies **only to real** numbers.

Suppose that we consider two real numbers **a** and **b**, and suppose also that **a** is greater than **b**. This fact can be indicated symbolically by $a > b$.

If **a** is less than **b**, we may write $a < b$. Note that the vertex of the symbol of inequality always points toward the smaller quantity.

Inequalities are of two kinds.

1) An **absolute inequality** is one which is true for all values of the letters involved.
2) A **conditional inequality** is one which is true for only certain values of the letters involved.

Illustration 1.

$(a - b)^2 > 0$ is an example of an absolute inequality. For regardless of whether $a > b$, or $a < b$, the square of the difference $(a - b)$ will be a positive number and hence be greater than zero.

Illustration 2.

$x - 4 > 0$ is an example of a conditional inequality. For if x is any negative value, the relationship is not true. If x is a positive number but less than 4, the relationship is not true. If $x = 4$, it is still not true. However, for all values of x which are greater than 4, $x - 4$ is a positive number and hence greater than zero.

97. Properties of Inequalities.

If $a > b$, and $c > d$, we say that the inequalities have the

same sense, since the signs of inequality point in the same direction.

If $a > b$, and $c < d$, we say that the inequalities have **opposite sense.**

The following properties are given here without proof.*

1) An inequality is **unchanged** in sense if the same number is added to or subtracted from both sides.

Illustration 1.

Since $5 > 3$

$$5 + 4 > 3 + 4.$$

Illustration 2.

Since $5 > 3$

$$5 - 4 > 3 - 4$$

$$\text{or} \quad 1 > -1.$$

2) An inequality is **unchanged** in sense if both sides are multiplied or divided by the same positive number.

Illustration 3.

Since $5 > 3$

$$\text{then} \quad 5 \cdot 2 > 3 \cdot 2.$$

Illustration 4.

Since $5 > 3$

$$\text{then} \quad 5/2 > 3/2.$$

3) The sense of an inequality is **changed** if both sides are multiplied or divided by the same negative number.

Illustration 5.

Since $5 > 3$

$$\text{then} \quad 5(-1) < 3(-1).$$

This may also be interpreted by saying that to change the signs of the members of an inequality, reverses the sense of an inequality.

Illustration 6.

Since $5 > 3$

$$\text{then} \quad 5/-2 < 3/-2.$$

---

* Proofs of these properties may be found in any college algebra.

4) The sense of an inequality is **unchanged** if the same positive root or power is taken of each side.

Illustration 7.

Since $5 > 3$

$$\text{then} \quad 5^2 > 3^2.$$

Illustration 8.

Since $25 > 16$

$$\text{then} \quad \sqrt{25} > \sqrt{16}$$

$$\text{or} \quad 5 > 4.$$

5) Terms may be transposed from one side to another in an inequality according to the rules which hold for equations.

Illustration 9.

If $7x - 5 > 3x + 4$

$$\text{then} \quad 7x - 3x - 5 > 4$$

$$\text{or} \quad 4x > 4 + 5, \quad \text{etc.}$$

6) If two inequalities have the same sense, their quotient does **not necessarily** have the same sense.

Illustration 10.

Given $8 > 6$ and $2 > 1$.

Then the quotient $8/2 \ngtr 6/1$ or $4 \ngtr 6$.
(The symbol $\ngtr$ means not greater than.)

98. Solution of Absolute Inequalities.

Illustration 1.

Show that $a^2 + 1/a^2 > 2$ if **a** is a positive number different from 1.

Solution: By the condition of the problem

$$a > 0 \text{ and } a \neq 1.$$

With these conditions on **a**, we see that

$$\text{either} \quad a - 1/a > 0 \quad \text{or} \quad 1/a - a > 0.$$

Squaring each of these: $a^2 - 2 + 1/a^2 > 0$

$$\text{or} \quad 1/a^2 - 2 + a^2 > 0.$$

In either case, add 2 to each side of the inequality and the expressions become the same, namely

$$a^2 + 1/a^2 > 2.$$

Illustration 2.

Show that $a^2 + b^2 > 2ab$ if **a** and **b** are any real numbers such that $a \neq b$.

Solution: Using Property 1, Sec. 97, subtract 2ab from both sides of the inequality, obtaining

$$a^2 - 2ab + b^2 > 0 \quad \text{or}$$
$$(a - b)^2 > 0.$$

Since $a \neq b$, then $(a - b)$ may be either a positive or negative quantity. But the square of any positive or negative quantity is always positive. Hence the last equation is true. All steps in the argument are reversible by Sec. 97, so that the given statement is true.

Illustration 3.

If **a** and **b** are positive numbers and $a > b$, show that $a^3 - b^3 > (a - b)^3$.

Solution: Write the given inequality in factor form, $(a - b)(a^2 + ab + b^2) > (a - b)(a - b)(a - b)$. Now since $a > b$, $a - b$ is a positive number and we may divide both sides by $(a - b)$ using Property 2, thus giving

$$a^2 + ab + b^2 > (a - b)^2 \quad \text{or}$$
$$a^2 + ab + b^2 > a^2 - 2ab + b^2.$$

Subtracting $a^2 + b^2$ using Property 1, we have

$$ab > -2ab \quad \text{or}$$
$$3ab > 0.$$

This last result is true; hence the given inequality is true.

99. Conditional Inequalities.

Since these inequalities are not true for all values of the numbers concerned, we shall want to determine the **value** or **range of values** for which the inequality holds. Two methods of solution will be shown, one algebraic, the other graphical.

Illustration 1.

For what values of **x** is $x - 3 > 0$?

Solution: By adding 3 to each side of the given inequality we have,

$$x > 3.$$

Illustration 2.

Determine values of x for which $x^2 - 2x - 8 > 0$.

Solution: This may be written

$$(x - 4)(x + 2) > 0.$$

This product will be positive if both factors are positive or both factors are negative. The inequality will not hold if one factor is positive and the other negative. Our problem then is to solve the separate inequalities: $x - 4 > 0$, $x + 2 > 0$, $x - 4 < 0$, $x + 2 < 0$. The first two of these are satisfied if $x > 4$ or $x > -2$; the second pair if $x < 4$ or $x < -2$. Hence we must conclude that both factors will be positive if $x > 4$, that both factors will be negative if $x < -2$. These two ranges of values therefore satisfy the given inequality.

Note that if x lies between -2 and $+4$ (expressed as $-2 < x < 4$), one factor is positive the other is negative, so that we can conclude that this is the interval for which the inequality **does not** hold.

In order to discuss the problem graphically we may set $y = x^2 - 2x - 8$, and draw the graph of this function as shown in Fig. 31.

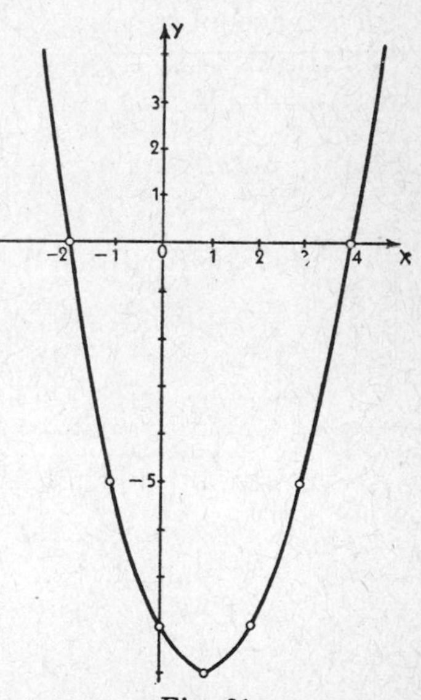

Fig. 31

The values of y for which the given inequality holds are those for which y is positive. From the graph we see that y is positive if $x > 4$ or $x < -2$. For all other values, namely $-2 < x < 4$, the graph lies below the x-axis and for this region y is negative. This means that from the graph we can readily find the solutions of $x^2 - 2x - 8 = 0$ as well as the solution of the given inequality.

Illustration 3.

Find the range of values of x for which

$$\frac{x - 3}{x + 2} > 0.$$

Solution: Since this fraction must be positive, we must have both numerator and denominator positive or have both numerator and denominator negative. But an expression changes sign from positive to negative by passing through the value zero. Hence we wish to note the values of x for which either the numerator or the denominator takes the value zero. Thus the changes of sign occur when $x = 3$ or $x = -2$ and for no other values. The whole x-axis is therefore divided into three ranges of values, those for which $x < -2$, $-2 < x < 3$, and $x > 3$.

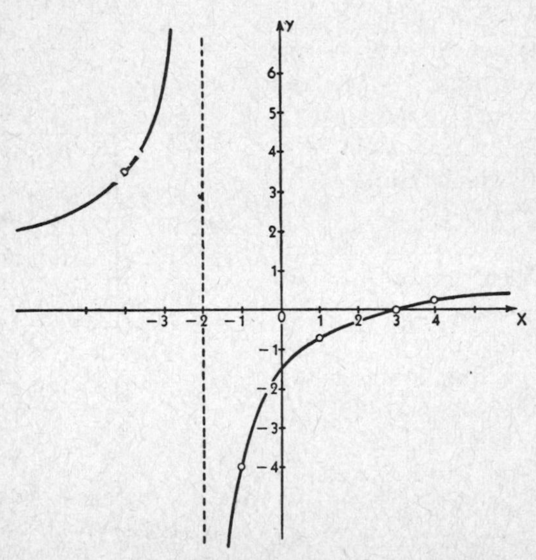

Fig. 32

If $x < -2$, both numerator and denominator are negative and the quotient is positive.

If $-2 < x < 3$, the numerator is negative, the denominator positive, and the fraction is negative. (These are the excluded cases for the given inequality.)

If $x > 3$, both numerator and denominator and the fraction are positive.

Hence we must conclude that the inequality is satisfied for all values of x which lie in the ranges for which $x < -2$ and $x > 3$.

The graphical interpretation for this problem is shown in Fig. 32. It is seen that the graph of $y = \frac{x-3}{x+2}$ lies above the x-axis for all values of $x > 3$ and all values of $x < -2$. The curve has an asymptote for $x = -2$, and a zero for $x = 3$, and these are the values for which the factors of the denominator and numerator each vanish.

Chapter XIV

LOGARITHMS

100. Irrational Exponents.

In Chapter IV, we defined the meaning of an exponent when the exponent was a positive or negative integer or fraction. Thus, $2^3 = 2 \cdot 2 \cdot 2 = 8$, and $4^{3/2}$ means the square root of the cube of 4, and equals 8. However, we have not as yet defined the meaning of a^x, where **a** is a positive number and **x** is an irrational number.

In order to generalize the use of exponents, we shall now **assume** that the laws of exponents hold for irrational as well as rational indices. By making this assumption we may now attach a meaning to such numbers as $3^{\sqrt{2}}$. By actually extracting square roots it can be shown that the $\sqrt{2}$ may be successively approximated as 1.4; 1.41; 1.414; 1.4142; etc., and thus we may approximate the value of $3^{\sqrt{2}}$ by writing successively:

$3^{\sqrt{2}}$ is approximately $3^{1.4} = 3^{1\frac{4}{10}}$
$3^{\sqrt{2}}$ is approximately $3^{1.41} = 3^{1\frac{41}{100}}$
$3^{\sqrt{2}}$ is approximately $3^{1.414} = 3^{1\frac{414}{1000}}$
etc.

Each approximation gives a more correct value of $3^{\sqrt{2}}$. Since $\sqrt{2}$ may be defined as a limit of a sequence of rational values, then $3^{\sqrt{2}}$ may be defined also in terms of limits. In general, it can be proved in advanced mathematics that if **x** approaches a limit, and if **a** is a positive number, a^x will have a limit. In this sense we define a^x as the limit approached when **x** is an irrational quantity. In other words, we are extending the laws of exponents to include irrational as well as rational indices.

101. Definition of a Logarithm.

The **logarithm** of a number M is the exponent **x**, indicating the power to which some base number **a** must be raised in order to equal M. The base number **a** must be positive and different from unity. This latter restriction is due to the fact that all powers of unity equal unity.

For example: $2^3 = 8$. Here the base is 2, the exponent (or logarithm) is 3, and the number is 8. We say then, that 3 is the logarithm of 8 when the base is 2, a fact which we usually symbolize by writing:

$$\log_2 8 = 3.$$

An alternative notation for defining a logarithm is as follows: If $a^x = M$, where $a > 0$, $a \neq 1$, then

$$x = \log_a M.$$

[This last statement is read, **x** is the logarithm of M to the base **a**.]

Certain restrictions have been placed on the values of **a**, **x**, and M. The base **a** has been restricted to positive numbers. If **a** could have negative values, then some positive numbers would have no real logarithm. For example, if $a = -2$, then $(-2)^x = 8$ cannot be satisfied by any real number **x**. If the base $a = 1$, then 1^x is always equal to 1 for all real values of **x** and hence no other numbers M could be expressed as $1^x = M$. The numbers M have been restricted to positive values. Yet in advanced courses this restriction may be removed and one may discuss the logarithms of negative and complex numbers.

With the definition of a logarithm as given, every positive number has a unique logarithm and every logarithm represents a unique number.

Illustration 1.

The following columns list by pairs, statements which are equivalent.

$2^3 = 8$	$\log_2 8 = 3$
$(\frac{1}{2})^4 = \frac{1}{16}$	$\log_{\frac{1}{2}} \frac{1}{16} = 4$
$16^{1/2} = 4$	$\log_{16} 4 = \frac{1}{2}$
$8^{5/3} = 32$	$\log_8 32 = \frac{5}{3}$
$2^{-3} = \frac{1}{8}$	$\log_2 (\frac{1}{8}) = -3$
$9^{-1/2} = \frac{1}{3}$	$\log_9 (\frac{1}{3}) = -\frac{1}{2}$

The left hand column gives the statement in **exponential form.** The right hand column gives the corresponding statement in **logarithmic form.** Opposite statements express the same fact, and the student should become sufficiently familiar with the two notations to be able to give the second statement whenever either of a pair is given.

102. Properties of Logarithms.

Since a logarithm is an exponent, we might suspect that the properties of logarithms reflect properties of exponents. This is true, and for that reason we should now recall the following three facts concerning exponents:

$$a^m \cdot a^n = a^{m+n}, \quad a^m \div a^n = a^{m-n}, \quad (a^m)^n = a^{m \cdot n}.$$

We now develop three theorems concerning logarithms, each of which corresponds to one of the above laws of exponents.

Theorem 1. The logarithm of the product of two numbers is equal to the sum of the logarithms of the numbers. In symbols: $\log_a M \cdot N = \log_a M + \log_a N$.

Proof: Let $a^x = M$ and $a^y = N$. (1)

Then $\log_a M = x$ and $\log_a N = y$. (2)

Now form the product of the two statements from (1).

$$a^x \cdot a^y = M \cdot N$$

or

$$a^{x+y} = M \cdot N. \qquad (3)$$

The corresponding logarithmic statement from (3) is

$$\log_a M \cdot N = x + y$$

But by (2) $\qquad = \log_a M + \log_a N.$

Theorem 2. The logarithm of the quotient of two numbers equals the logarithm of the numerator minus the logarithm of the denominator.

i.e., $\log_a M/N = \log_a M - \log_a N$.

Proof: From (1) above, form the quotient

$$\frac{a^x}{a^y} = \frac{M}{N} = a^{x-y}.$$

Then $\log_a M/N = x - y$

$$= \log_a M - \log_a N.$$

Theorem 3. The logarithm of the kth power of a number equals **k** times the logarithm of the number.

i.e., $\log_a M^k = k \cdot \log_a M$.

Proof: Let $a^x = M$ and raise both sides to the kth power. Then

$$a^{kx} = M^k.$$

Writing this in logarithmic form gives

$$\log_a M^k = kx = k \cdot \log_a M.$$

The index **k** may be any constant. If **k** is a fraction, we have from the relationships between fractional exponents and roots, a means of expressing the logarithm of the root of a number. Thus,

$$\log_a \sqrt{M} = \log_a M^{1/2} = \tfrac{1}{2} \log_a M.$$
$$\log_a \sqrt[3]{M} = \log_a M^{1/3} = \tfrac{1}{3} \log_a M, \text{ etc.}$$

In general, $\mathbf{\log_a \sqrt[k]{M} = \log_a M^{1/k} = \frac{1}{k} \log_a M.}$

Illustration 1.

Express $\log_a \frac{(571)^3(1.87)}{32.6}$ as an algebraic sum in terms of the logarithms of the factors.

Solution: By Theorems 1 and 2

$$\log_a \frac{(571)^3(1.87)}{32.6} = \log_a (571)^3 + \log_a 1.87 - \log_a 32.6$$

By Theorem 3,

$$= 3 \log_a 571 + \log_a 1.87 - \log_a 32.6.$$

Illustration 2.

Express $\log_a \sqrt{75}$ in terms of the logarithms of prime numbers.

Solution: Since $75 = 5^2 \cdot 3$

$$\log_a \sqrt{75} = \tfrac{1}{2} \log_a 75 = \tfrac{1}{2} \log_a 5^2 \cdot 3 = \tfrac{1}{2} [\log_a 5^2 + \log_a 3]$$
$$= \tfrac{1}{2}[2 \cdot \log_a 5 + \log_a 3] = \log_a 5 + \tfrac{1}{2} \log_a 3.$$

Illustration 3.

Express $2 \log 3 + \log 14 + \log 5 - \log 3 - \log 2$ as a single logarithm.

Solution: Using the theorems, we have

$$2 \log 3 + \log 14 + \log 5 - \log 3 - \log 2 =$$
$$\log 3^2 + \log 14 + \log 5 - \log 3 - \log 2 =$$
$$\log \frac{3^2 \cdot 14 \cdot 5}{3 \cdot 2} = \log 3 \cdot 7 \cdot 5 = \log 105.$$

Illustration 4.

Show that Theorem 1 may be extended so that

$$\log_a L \cdot M \cdot N = \log_a L + \log_a M + \log_a N.$$

Solution: Let $LM = Q$, then by Theorem 1

$$\begin{aligned}\log_a QN &= \log_a Q + \log_a N.\\ &= \log_a LM + \log_a N\\ &= \log_a L + \log_a M + \log_a N.\end{aligned}$$

103. Common Logarithms.

Since our number system is built upon 10 as a basis, we find it most convenient to use 10 as the base of a system of logarithms. **Common logarithms** are those for which the base 10 is used. Unless otherwise specified we shall use base $a = 10$, and the base will not be indicated. Thus $\log_{10} 17$ will simply be written log 17.

In order to have an understanding of the use of logarithms it is convenient for us to study the parallel columns of the data which follow.

Powers of 10.	Logarithmic notation.
.	.
.	.
.	.
$10^3 = 1000$	$\log 1000 = 3$
$10^2 = 100$	$\log 100 = 2$
$10^1 = 10$	$\log 10 = 1$
$10^0 = 1$	$\log 1 = 0$
$10^{-1} = \frac{1}{10} = .1$	$\log .1 = -1$
$10^{-2} = \frac{1}{100} = .01$	$\log .01 = -2$
$10^{-3} = \frac{1}{1000} = .001$	$\log .001 = -3$
.	.
.	.
etc.	etc.

If we confine our attention to integral powers of ten, the logarithms may be obtained from the above chart. But suppose that we wish to find log 253.

Since $10^2 < 253 < 10^3$,

then $2 < \log 253 < 3$.

If we let $x = \log 253$, then $10^x = 253$, and we may write

$2 < x < 3$ or

$x = 2 +$ (some decimal value),

the added quantity being less than 1. This decimal part is called the **mantissa** of the logarithm. The integral part (such as 2 above) is called the **characteristic.** The mantissa can be found from computed tables such as those shown on pages 126 and 127. Using these tables we find the amount to be added.

The table gives the mantissas to four significant figures for three figure numbers.*

The first two digits of the number are given in the left hand column under N, and the third digit is arranged across the top of the page. Thus, to find the mantissa for 253, look under column N for 25 and across under the column labeled 3. Here one finds the mantissa for 253 given as 4031. Consequently,

$$\log 253 = 2.4031.$$

Numbers like 00253, .0253, 253, 253000, and 0.253 are said to have the **same sequence** of digits. In ascertaining the sequence of digits, initial and final zeros are disregarded.

Such numbers, however, have different logarithms. The mantissa for each of these numbers is the same, but the characteristic is different. Thus, 0.0253 lies between 0.01 and 0.1; hence its logarithm lies between -2 and -1. A mantissa is **always** a positive quantity. In order to avoid confusion in computation, it is customary to write a negative characteristic like -2, as $8 - 10$.

With this convention, we may now write

$$\log 0.0253 = 8.4031 - 10.$$

In a similar way,

$$\log 0.00253 = 7.4031 - 10$$
$$\log 253000 = 5.4031.$$

The mantissa is the same for all numbers which have the same sequence of digits. The mantissa is always a positive decimal fraction.

---

* Five place, seven place, and even twenty place tables are available for computing purposes. The four place table, however, is sufficient for a study of the theory and use of logarithms.

LOGARITHMS

N	0	1	2	3	4	5	6	7	8	9
10	0000	0043	0086	0128	0170	0212	0253	0294	0334	0374
11	0414	0453	0492	0531	0569	0607	0645	0682	0719	0755
12	0792	0828	0864	0899	0934	0969	1004	1038	1072	1106
13	1139	1173	1206	1239	1271	1303	1335	1367	1399	1430
14	1461	1492	1523	1553	1584	1614	1644	1673	1703	1732
15	1761	1790	1818	1847	1875	1903	1931	1959	1987	2014
16	2041	2068	2095	2122	2148	2175	2201	2227	2253	2279
17	2304	2330	2355	2380	2405	2430	2455	2480	2504	2529
18	2553	2577	2601	2625	2648	2672	2695	2718	2742	2765
19	2788	2810	2833	2856	2878	2900	2923	2945	2967	2989
20	3010	3032	3054	3075	3096	3118	3139	3160	3181	3201
21	3222	3243	3263	3284	3304	3324	3345	3365	3385	3404
22	3424	3444	3464	3483	3502	3522	3541	3560	3579	3598
23	3617	3636	3655	3674	3692	3711	3729	3747	3766	3784
24	3802	3820	3838	3856	3874	3892	3909	3927	3945	3962
25	3979	3997	4014	4031	4048	4065	4082	4099	4116	4133
26	4150	4166	4183	4200	4216	4232	4249	4265	4281	4298
27	4314	4330	4346	4362	4378	4393	4409	4425	4440	4456
28	4472	4487	4502	4518	4533	4548	4564	4579	4594	4609
29	4624	4639	4654	4669	4683	4698	4713	4728	4742	4757
30	4771	4786	4800	4814	4829	4843	4857	4871	4886	4900
31	4914	4928	4942	4955	4969	4983	4997	5011	5024	5038
32	5051	5065	5079	5092	5105	5119	5132	5145	5159	5172
33	5185	5198	5211	5224	5237	5250	5263	5276	5289	5302
34	5315	5328	5340	5353	5366	5378	5391	5403	5416	5428
35	5441	5453	5465	5478	5490	5502	5514	5527	5539	5551
36	5563	5575	5587	5599	5611	5623	5635	5647	5658	5670
37	5682	5694	5705	5717	5729	5740	5752	5763	5775	5786
38	5798	5809	5821	5832	5843	5855	5866	5877	5888	5899
39	5911	5922	5933	5944	5955	5966	5977	5988	5999	6010
40	6021	6031	6042	6053	6064	6075	6085	6096	6107	6117
41	6128	6138	6149	6160	6170	6180	6191	6201	6212	6222
42	6232	6243	6253	6263	6274	6284	6294	6304	6314	6325
43	6335	6345	6355	6365	6375	6385	6395	6405	6415	6425
44	6435	6444	6454	6464	6474	6484	6493	6503	6513	6522
45	6532	6542	6551	6561	6571	6580	6590	6599	6609	6618
46	6628	6637	6646	6656	6665	6675	6684	6693	6702	6712
47	6721	6730	6739	6749	6758	6767	6776	6785	6794	6803
48	6812	6821	6830	6839	6848	6857	6866	6875	6884	6893
49	6902	6911	6920	6928	6937	6946	6955	6964	6972	6981
50	6990	6998	7007	7016	7024	7033	7042	7050	7059	7067
51	7076	7084	7093	7101	7110	7118	7126	7135	7143	7152
52	7160	7168	7177	7185	7193	7202	7210	7218	7226	7235
53	7243	7251	7259	7267	7275	7284	7292	7300	7308	7316
54	7324	7332	7340	7348	7356	7364	7372	7380	7388	7396

LOGARITHMS

N	0	1	2	3	4	5	6	7	8	9
55	7404	7412	7419	7427	7435	7443	7451	7459	7466	7474
56	7482	7490	7497	7505	7513	7520	7528	7536	7543	7551
57	7559	7566	7574	7582	7589	7597	7604	7612	7619	7627
58	7634	7642	7649	7657	7664	7672	7679	7686	7694	7701
59	7709	7716	7723	7731	7738	7745	7752	7760	7767	7774
60	7782	7789	7796	7803	7810	7818	7825	7832	7839	7846
61	7853	7860	7868	7875	7882	7889	7896	7903	7910	7917
62	7924	7931	7938	7945	7952	7959	7966	7973	7980	7987
63	7993	8000	8007	8014	8021	8028	8035	8041	8048	8055
64	8062	8069	8075	8082	8089	8096	8102	8109	8116	8122
65	8129	8136	8142	8149	8156	8162	8169	8176	8182	8189
66	8195	8202	8209	8215	8222	8228	8235	8241	8248	8254
67	8261	8267	8274	8280	8287	8293	8299	8306	8312	8319
68	8325	8331	8338	8344	8351	8357	8363	8370	8376	8382
69	8388	8395	8401	8407	8414	8420	8426	8432	8439	8445
70	8451	8457	8463	8470	8476	8482	8488	8494	8500	8506
71	8513	8519	8525	8531	8537	8543	8549	8555	8561	8567
72	8573	8579	8585	8591	8597	8603	8609	8615	8621	8627
73	8633	8639	8645	8651	8657	8663	8669	8675	8681	8686
74	8692	8698	8704	8710	8716	8722	8727	8733	8739	8745
75	8751	8756	8762	8768	8774	8779	8785	8791	8797	8802
76	8808	8814	8820	8825	8831	8837	8842	8848	8854	8859
77	8865	8871	8876	8882	8887	8893	8899	8904	8910	8915
78	8921	8927	8932	8938	8943	8949	8954	8960	8965	8971
79	8976	8982	8987	8993	8998	9004	9009	9015	9020	9025
80	9031	9036	9042	9047	9053	9058	9063	9069	9074	9079
81	9085	9090	9096	9101	9106	9112	9117	9122	9128	9133
82	9138	9143	9149	9154	9159	9165	9170	9175	9180	9186
83	9191	9196	9201	9206	9212	9217	9222	9227	9232	9238
84	9243	9248	9253	9258	9263	9269	9274	9279	9284	9289
85	9294	9299	9304	9309	9315	9320	9325	9330	9335	9340
86	9345	9350	9355	9360	9365	9370	9375	9380	9385	9390
87	9395	9400	9405	9410	9415	9420	9425	9430	9435	9440
88	9445	9450	9455	9460	9465	9469	9474	9479	9484	9489
89	9494	9499	9504	9509	9513	9518	9523	9528	9533	9538
90	9542	9547	9552	9557	9562	9566	9571	9576	9581	9586
91	9590	9595	9600	9605	9609	9614	9619	9624	9628	9633
92	9638	9643	9647	9652	9657	9661	9666	9671	9675	9680
93	9685	9689	9694	9699	9703	9708	9713	9717	9722	9727
94	9731	9736	9741	9745	9750	9754	9759	9763	9768	9773
95	9777	9782	9786	9791	9795	9800	9805	9809	9814	9818
96	9823	9827	9832	9836	9841	9845	9850	9854	9859	9863
97	9868	9872	9877	9881	9886	9890	9894	9899	9903	9908
98	9912	9917	9921	9926	9930	9934	9939	9943	9948	9952
99	9956	9961	9965	9969	9974	9978	9983	9987	9991	9996

104. Rules for Characteristic.

There are two rules for determining the characteristic of a logarithm. They are found by inspection from the chart of powers of ten in Sec. 103. Any number which lies between 100 and 1000 will have a characteristic equal to 2, and all numbers in this interval have three digits to the left of the decimal point. All numbers between 10 and 100 will have characteristic 1, and such numbers have two digits to the left of the decimal point. The general rule may be stated as follows:

Rule 1.

If the number N is greater than 1, then its characteristic is one less than the number of digits to the left of the decimal point.

If the number lies in the interval from zero to 1, we may find its characteristic according to,

Rule 2.

If the number N is less than unity, and if the first digit which is not zero occurs in the kth decimal place, the characteristic is $-k$.

Illustration 1.

Write the characteristic of the logarithm for each of the following numbers: 287, 3.47, 9640, 0.124, 0.00820, 0.000703.

Solution: Using the above rules we have,

287	has characteristic	2;
3.4	has characteristic	0;
9640	has characteristic	3;
0.124	has characteristic	-1 or $9 - 10$;
0.00820	has characteristic	-3 or $7 - 10$;
0.000703	has characteristic	-4 or $6 - 10$.

Illustration 2.

Write the logarithms of each of the numbers in Illustration 1 (use tables pp. 126–127).

Solution:

$$\begin{aligned} \log 287 &= 2.4579 \\ \log 3.4 &= 0.5315 \\ \log 9640 &= 3.9841 \\ \log 0.124 &= 9.0934 - 10 \\ \log 0.0082 &= 7.9138 - 10 \\ \log 0.000703 &= 6.8470 - 10. \end{aligned}$$

105. Use of Tables.

The tables of mantissas may be used for two purposes. One may find from them the logarithm of a given number, or find the number corresponding to a given logarithm.

a) To find the logarithm of a given number.

Illustration 1.

Find log 29.6.

Solution: From the table on pp. 126–127, under column N, find 29. Opposite 29 and under the column labeled 6 one finds the mantissa 4713.

Hence, $\log 29.6 = 1.4713$ since its characteristic is 1.

b) To find the number, given its logarithm.

Illustration 2.

Find the number whose logarithm is $7.5843 - 10$.

Solution: We have given $\log N = 7.5843 - 10$. We must locate this mantissa in the table, find the number which corresponds to it, and place the decimal point in accord with rule 2 for characteristics. In the table, pp. 126–127, we find that the mantissa 5843 corresponds to the digit sequence 384. Since the characteristic is -3, we must have two zeros following the decimal point and the first non-zero digit in the third place.

Hence $N = 0.00384$.

106. Interpolation.

Frequently one wishes to find the mantissa for a number of four or five digits. Since these are not given directly by the table, we must devise a means for estimating the value of the mantissa. The process by which this is done is called **interpolation,** since it involves placing values **between** known values from the tables.

We **assume** that for small changes in the number, there will be a corresponding small but proportional change in the mantissa. A few illustrations should make the method clear.

Illustration 1.

Find log 752.3.

Solution: Since this number does not appear directly in the tables, we take the two values closest to it which do appear. Using the tables, we find,

$$\left.\begin{array}{l}\left.\begin{array}{l}\log 752.0 = 2.8762 \\ \log 752.3 = \quad ?\end{array}\right\} x \\ \log 753.0 = 2.8768\end{array}\right\} .0006.$$

The difference between the mantissas is .0006. We assume the difference between the mantissa given and the mantissa we want is some value x. The number 7523 lies $\frac{3}{10}$ of the way from 7520 to 7530. Hence we may write,

$$x/.0006 = \tfrac{3}{10} \quad \text{or} \quad x = .00018 = .0002.$$

Adding .0002 to the mantissa .8762 we have .8764 and may now write

$$\log 752.3 = 2.8764.$$

Note: We keep all mantissas to 4 decimals. Hence .00018 was called .0002. If the amount added is a value like .00075, we call it .0008 or .0007 according as the amount added makes the final digit even. This is an old practice known as the **computer's** rule.

Illustration 2.

Find N if log N = 8.3754 − 10.

Solution: This mantissa does not lie in the table. We take the tabular values on each side of it and write the numbers corresponding to them:

$$10\left\{\begin{array}{l}\left.\begin{array}{l}2370 \\ \ \ ?\end{array}\right\} x \\ \\ 2380\end{array}\right. \quad \left.\begin{array}{l}\left.\begin{array}{l}\text{corresponds to } .3747 \\ \text{corresponds to } .3754\end{array}\right\} .0007 \\ \\ \text{corresponds to } .3766\end{array}\right\} .0019.$$

The correction to be made on 2370 is called x. Now 3754 lies $\frac{7}{19}$ of the way between the given mantissas. We must determine a value x which lies $\frac{7}{19}$ of the way between 2370 and 2380.

Hence $x/10 = \frac{7}{19}$ or $x = \frac{70}{19} = 3\frac{13}{19}$ (called 4).

The digits of the number are then 2374.

Therefore N = 0.02374.

The number corresponding to a given logarithm is called an **antilogarithm.**

107. Computation Using Logarithms.

Illustration 1.

Find N if $N = \frac{(278)(.0034)}{17.6}$.

Solution: The logarithmic statement is

$$\log N = \log \frac{(278)(.0034)}{17.6}$$

$$= \log 278 + \log .0034 - \log 17.6.$$

$$\begin{aligned}
\text{Now, } \log 278 &= 2.4440 \\
\log .0034 &= \underline{7.5315 - 10} \\
\text{Adding} \quad & 9.9755 - 10 \\
\text{Also, } \log 17.6 &= \underline{1.2455} \\
\text{Subtracting, } \log N &= 8.7300 - 10 \\
\text{Therefore, } N &= .0537.
\end{aligned}$$

Illustration 2.

Find N if, $N = \sqrt[3]{\frac{(63.7)(2.46)}{(13.2)^2}}$.

Solution: $\log N = \frac{1}{3}[\log 63.7 + \log 2.46 - 2 \log 13.2]$.

$$\begin{aligned}
\text{Now } \log 63.7 &= 1.8041 \\
\log 2.46 &= \underline{0.3909} \\
\text{Adding} \quad & 2.1950 \qquad (1) \\
2 \log 13.2 &= 2.2412. \qquad (2)
\end{aligned}$$

Before (2) can be subtracted from (1) we must write (1)

$$\begin{aligned}
\text{as:} \quad & 12.1950 - 10 \\
\text{Repeat (2)} \quad & \underline{2.2412} \\
\text{Subtract} \quad & 9.9538 - 10. \qquad (3)
\end{aligned}$$

Before taking $\frac{1}{3}$ of this value we write (3) in the form 29.9538 − 30 so that when it is divided by three, we shall have the 9 − 10 form for the characteristic. If (3) is divided as it stands, one gets

$$\frac{9 \cdot 9538 - 10}{3} = 3.3179 - 3.3333$$

and one must actually perform this subtraction to find the logarithm.

$$\text{However, } \log N = \tfrac{1}{3}(29.9538 - 30) = 9.9846 - 10$$
$$\text{and } N = .9652.$$

Illustration 3.

Solve for x if, $x = \dfrac{\log .016}{\log 5.2}$.

Solution: Note, that this is the quotient of two logarithms, and **not** the logarithm of a quotient covered by Theorem 2. Consequently we have upon finding the logarithms,

$$x = \frac{8.2041 - 10}{0.7160}. \tag{1}$$

But now our problem presents another difficulty. The logarithm $8.2041 - 10$ or $-2 + .2041$ represents a number composed of -2 plus a positive mantissa. On a linear scale one can interpret this as a number obtained by adding .2041 to -2 on the scale.

One must translate this number

$-2 + 0.2041$ or -1.7959

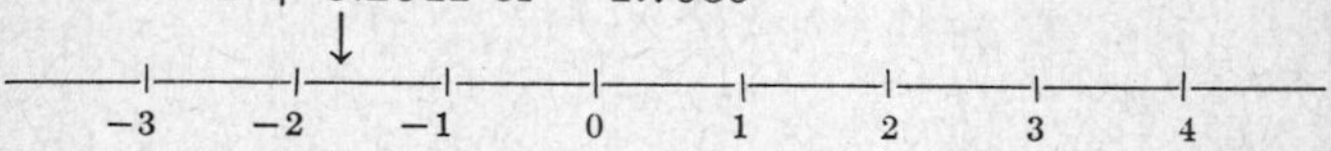

↓

−3 −2 −1 0 1 2 3 4

into an ordinary negative number by adding .2041 to -2, thus obtaining -1.7959.

Equation (1) may now be written as

$$x = \frac{-1.7959}{.7160} \text{ and by actual division}$$

$$x = -2.508.$$

The division above can also be accomplished by use of logarithms. One disregards the negative sign and proceeds to compute the value of x from $\log x = \log \dfrac{1.7959}{.7160}$ $= \log 1.7959 - \log .7160$. The final result is assigned the negative sign since the algebraic result is itself negative.

108. Cologarithms.

The **cologarithm** of a number is defined as the logarithm of the reciprocal of the number. Thus

$$\begin{aligned} \text{Colog } N &= \log 1/N = \log 1 - \log N \\ &= 0 - \log N = -\log N \\ &= 10.0000 - 10 - \log N. \end{aligned}$$

Illustration 1.

Find colog 16.4.

Solution:

$$\begin{array}{ll} & \text{Colog } 16.4 = \log 1 - \log 16.4 \\ & \qquad\qquad\;\;\; = 10.0000 - 10 \\ \text{minus} & \qquad\qquad\quad\;\; \underline{1.2148\qquad\;\;}. \\ \text{colog } 16.4 & \qquad\qquad\quad\;\; 8.7852 - 10 \end{array}$$

Computation may be carried on using cologarithms instead of logarithms for cases such as the following:

Illustration 2.

$$\begin{aligned} \log \frac{173}{5.4} &= \log 173 - \log 5.4 \\ &= \log 173 + \text{colog } 5.4. \end{aligned}$$

109. Exponential and Logarithmic Equations.

An **exponential equation** is one in which the unknown occurs as an exponent.

Illustration 1.

$4^x = 64$ and $2^{x^2 - 3x + 1} = 8$ are exponential equations.

A **logarithmic equation** is one in which the logarithm involves the unknown quantity.

Illustration 2.

$\log (x^2 - 5) = 2 \log x$, and $\log x + 5 = 10$ are examples of logarithmic equations.

The following illustrations are given to show methods of solving exponential and logarithmic equations.

Illustration 3.

Solve $3^{x-1} = 2^x$ for x.

Solution: Take the logarithm of both numbers and

$$\log 3^{x-1} = \log 2^x$$
$$(x - 1)\log 3 = x \log 2$$
$$x \log 3 - \log 3 = x \log 2$$
$$x(\log 3 - \log 2) = \log 3$$
$$x = \frac{\log 3}{\log 3 - \log 2} = \frac{0.4771}{(0.4771 - 0.3010)}$$
$$= \frac{0.4771}{0.1761} = 2.709+$$

where the result is given to four significant figures.

Illustration 4.

Solve, $\log x - \log (x - 9) = 1$.

Solution: This may be written,

$$\log \frac{x}{x - 9} = 1.$$

Writing this in exponential form

$$\frac{x}{x-9} = 10^1 = 10$$
$$x = 10x - 90$$
$$9x = 90$$
$$x = 10.$$

110. The Graphs of Exponential and Logarithmic Equations.

Consider first the graph of $y = 2^x$. Construct a chart of values as below.

x	−3	−2	−1	0	1	2	3	4	
y	$\frac{1}{8}$	$\frac{1}{4}$	$\frac{1}{2}$	1	2	4	8	16	

The curve obtained from the points of this chart is shown in Fig. 33. It is characteristic of the appearance of all exponential functions for which $y = a^x$, where $a > 0$. The exponential curve $y = 2^{-x}$ is shown in the same figure as a **dotted** line curve. It is characteristic of the shape of all curves of the form $y = a^{-x}$ where $a > 0$.

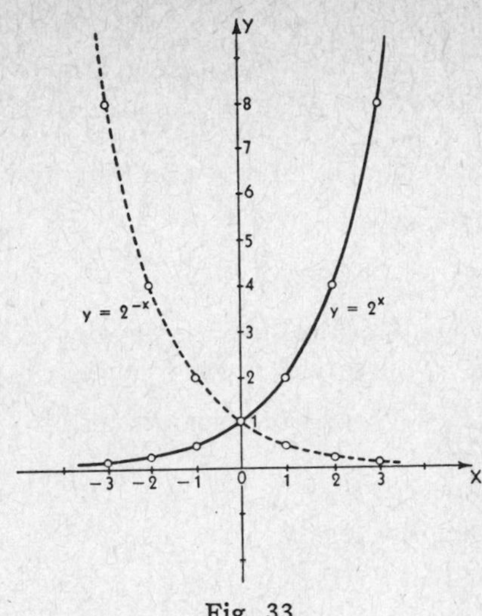

Fig. 33

Now consider the graph of the curve $y = \log_2 x$. Construct the chart:

x	$\frac{1}{8}$	$\frac{1}{4}$	$\frac{1}{2}$	1	2	4	8	16	
y	−3	−2	−1	0	1	2	3	4	

The graph of the data of this chart is shown in Fig. 34. The curve $y = \log_2 x$ is characteristic of the shape of all functions of the type $y = \log_a x$, where $a > 1$. If $a < 1$, the graph of such functions will look like the dotted curve in Fig. 34, which shows the graph of $y = \log_{\frac{1}{2}} x$.

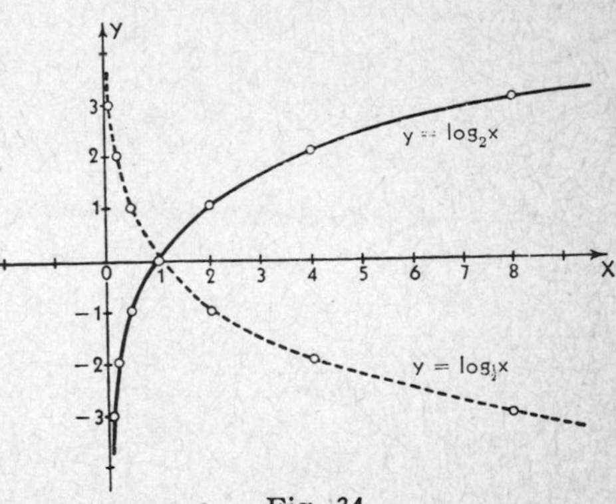

Fig. 34

111. Change of Base.

One of the fundamental uses of the theory of logarithms and exponentials is to express a given number in terms of different bases.

How this may be done is shown in the following illustration.

Illustration 1.

Express 17 as a power of 3.

Solution: This problem is equivalent to saying, find the $\log_3 17$. It is solved in the following way:

Let $3^x = 17$, and take logarithms of both sides to base 10; then

$$\log 3^x = \log 17,$$
$$x \log 3 = \log 17,$$
$$x = \frac{\log 17}{\log 3} = \frac{1.2304}{0.4771} = 2.58 \text{ (approx.)}.$$

With two place significant figures we may then say that $17 = 3^{2.58}$.

The general formula for change of base is

$$\log_b N = \frac{\log_a N}{\log_a b}$$

and expresses the logarithm of N to base **b** if a table with base **a** is given.

This relationship is proved as follows.

Suppose the number N is expressed in two different bases,

$$N = a^x = b^y, \text{ so that}$$
$$a^x = b^y.$$

Take the logarithm of both sides to base **a**,

$$\log_a a^x = \log_a b^y \quad \text{or}$$
$$x \log_a a = y \log_a b, \quad \text{and since } \log_a a = 1,$$
$$x = y \log_a b.$$

But $x = \log_a N$ and $y = \log_b N$, so that

$$\log_a N = \log_b N \cdot \log_a b \quad \text{or}$$
$$\log_b N = \frac{\log_a N}{\log_a b}.$$

There is an irrational number $e = 2.7182 \ldots$, called the

natural base, which plays an important role in the theory of logarithms. Its value and definition are usually given in a course in calculus. However, logarithms to the base **e** may be computed from a table of base 10 by means of

$$\log_e N = \frac{\log_{10} N}{\log_{10} e}$$

$$\log_{10} e = .4343 \ldots, \quad 1/\log_{10} e = 2.3026 \ldots.$$

The $\log_{10} e$ is known as the **modulus** of common logarithms with respect to natural logarithms. Its value may be used as a conversion factor to change from base 10 to base **e.**

Illustration 2.

Express $\log_4 16$ in terms of the logarithm with base 2.

Solution: From the formula for change of base, one has

$$\log_4 16 = \frac{\log_2 16}{\log_2 4} = \frac{4}{2} = 2$$

as could have been stated at once from the definition.

The student need not remember the formula for change of base, if problems are solved by the method of the first illustration in this article.

CHAPTER XV

PROGRESSIONS

112. Introduction.

A **progression** is a sequence of numbers formed according to some law. We shall consider three types in this chapter: arithmetic progressions, geometric progressions, and harmonic progressions.

113. Arithmetic Progressions.

Definition: An **arithmetic progression** is a sequence of numbers each of which differs from the one which precedes it by a constant amount called the **common difference.**

If the terms of the progression are in an increasing order, we shall consider the common difference to be **positive.** An example of such a progression is,

$$3, 6, 9, 12, \ldots$$

in which the common difference is $+3$.

If the terms appear in decreasing order, we shall consider the common difference to be **negative.** An example of this type is,

$$22, 20, 18, 16, \ldots$$

in which the common difference is -2.

We shall also adopt the following notation in our discussion of arithmetic progressions.

Let a, represent the first term of the progression,
d, the common difference,
n, the number of terms in the progression,
l, the last term (n^{th} term) under consideration,
s, the sum of n terms of the progression.

We shall call these five letters the **indices** of the progression.

114. Relationships among the Indices (A.P.).

We wish now to consider relationships among these five indices which will enable us to determine the numerical values of all of them provided any three of them are known for a given progression.

We note that if **a** is the first term and **d** is the common difference, then

$$\begin{aligned} a + d &= \text{2nd term}, \\ a + 2d &= \text{3rd term}, \\ a + 3d &= \text{4th term}, \\ &\text{etc.} \end{aligned}$$

In general the nth or last term l is given by

$$l = a + (n - 1)d. \tag{1}$$

That the coefficient of **d** is (n − **1**) is seen by inspection. For we note above that the coefficient of **d** for any particular term is numerically **one less than** the number of the term.

We now assume that we have an arithmetic progression consisting of **n** terms, with first term **a**, and last term l. The next to last term may be written as $l - d$, the term preceding it as $l - 2d$, etc. The sum of these **n** terms may be written as

$$\begin{aligned} s = a + (a + d) + (a + 2d) + \dots \\ + (l - 2d) + (l - d) + l, \end{aligned} \tag{2}$$

where the sequence of dots indicates that some of the intermediate terms are not shown.

If the progression is rewritten in the reverse order, one has,

$$\begin{aligned} s = l + (l - d) + (l - 2d) + \dots \\ + (a + 2d) + (a + d) + a. \end{aligned} \tag{3}$$

Upon adding the expressions (2) and (3) one obtains

$$\begin{aligned} 2s = (a + l) + (a + l) + (a + l) + \dots \\ + (a + l) + (a + l) + (a + l). \end{aligned}$$

Since the progression contained **n** terms, there will be **n** of these quantities $(a + l)$; so one may write,

$$2s = n(a + l) \qquad \text{or}$$

$$s = \frac{n}{2}(a + l). \tag{4}$$

Equations (1) and (4) make it possible for us to find values for all of the five indices whenever any three of them are known.

Illustration 1.

Find the sum of all the numbers between 1 and 100 which are divisible by 3.

Solution: These numbers form an arithmetic progression with first term $a = 3$, $d = 3$ and $l = 99$. Using these values in equation (1), we have

$$99 = 3 + (n - 1)3 \quad \text{or}$$
$$99 = 3 + 3n - 3$$
$$99 = 3n$$
$$n = 33.$$

The sum may now be obtained from equation (4).

$$s = n\,(a + l)/2 = 33\,(3 + 99)/2 = 33\,(102)/2 = 33 \cdot 51 = 1683.$$

Illustration 2.

Find **a** and **n**, if given $d = 4$, $s = 7$, and $l = 13$.

Solution: Since $l = a + (n - 1)d$

$$13 = a + (n - 1)4$$
$$a = 17 - 4n.$$

But $s = n/2(a + l) = n/2(17 - 4n + 13)$

$$7 = n(15 - 2n)$$
$$2n^2 - 15n + 7 = 0$$
$$\text{or } (2n - 1)(n - 7) = 0.$$

But since n must be a positive integer,

$$n = 7$$

and $a = 17 - 4n = 17 - 28 = -11.$

115. Geometric Progressions.

A **geometric progression** is a sequence of numbers such that any term after the first is obtained from the preceding term by multiplying it by a fixed number called the common ratio. The sequence

$$-4, -2, -1, -\tfrac{1}{2}, -\tfrac{1}{4}, \ldots$$

is a geometric progression with ratio $\frac{1}{2}$.

116. Relationships among the Indices (G.P.)

For a geometric progression, let **a** be the first term, **r** be the common ratio, l the nth term, s the sum of n terms.

Then ar = second term,

ar^2 = third term,

$\vdots \quad \vdots \quad \vdots$

and $l = ar^{n-1} = n^{th}$ term. (1)

The sum of these n terms can be written,

$$s = a + ar + ar^2 + ar^3 + \ldots + ar^{n-1}. \quad (2)$$

Now multiply each term of (2) by r, and

$$rs = ar + ar^2 + ar^3 + ar^4 + \ldots ar^{n-1} + ar^n. \quad (3)$$

Subtracting (3) from (2) gives

$$s - rs = a - ar^n \quad \text{or}$$

$$s = \frac{a(1 - r^n)}{1 - r}, \text{ providing } r \neq 1. \quad (4)$$

Since $l = ar^{n-1}$ this may also be written,

$$s = \frac{a - rl}{1 - r}, \; r \neq 1. \quad (5)$$

These formulae enable us to find all of the indices for a G.P., whenever any three of them are known.

Illustration 1.

In a G.P., $a = 1$, $n = 3$, and $s = 57$. Find values of r satisfying these conditions and write out a few of the terms of the progressions which are determined.

Solution: By formula (4) $57 = \frac{1 - r^3}{1 - r}$, or

$$57 = 1 + r + r^2$$

$$r^2 + r - 56 = 0$$

$$(r + 8)(r - 7) = 0$$

$$\text{so that } r = 7 \text{ or } r = -8.$$

Hence, there are two progressions determined:

If $r = 7$, the progression is $1, 7, 49, \ldots,$

If $r = -8$, the progression is $1, -8, 64, \ldots.$

117. Harmonic Progressions.

A **harmonic progression** is a sequence of numbers such that the sequence formed from the reciprocals is an arithmetic progression.

Illustration 1.

Show that $\frac{1}{3}$, $\frac{1}{5}$, $\frac{1}{7}$, etc., is a harmonic progression.

Solution: Forming the sequence of the reciprocals of these numbers gives,

$$3, 5, 7, \text{etc.}$$

This last sequence is an arithmetic progression; hence, the given sequence is harmonic.

The indices for a harmonic progression are found by first writing the corresponding arithmetic progression, then translating this information according to the definition of harmonic progression. No general formula is given for the sum, and there is no common difference or common ratio.

118. Means.

The terms of a progression between any two given terms are called the **means** between those two terms.

If the progression is **arithmetic,** one applies the name **arithmetic means**; if the sequence is **geometric** or **harmonic,** one applies the terminology **geometric** or **harmonic means.**

The problem of inserting means does not present any new ideas in connection with progressions. How means are inserted is shown below.

Illustration 1.

Insert three harmonic means between $-\frac{3}{2}$ and $\frac{3}{10}$.

Solution: One must first insert three arithmetic means between $-\frac{2}{3}$ and $\frac{10}{3}$. Thus we see that we are asked to form an arithmetic progression consisting of 5 terms (the 2 given terms and the 3 means), whose first term $a = -\frac{2}{3}$ and last term $l = \frac{10}{3}$. From $l = a + (n - 1)d$, we have,

$$\frac{10}{3} = -\frac{2}{3} + 4d, \quad \text{or}$$
$$10 = -2 + 12d$$
$$12 = 12d$$
$$d = 1.$$

Hence the three arithmetic means are,

$$-\frac{2}{3} + 1 = \frac{1}{3}, \frac{1}{3} + 1 = \frac{4}{3}, \frac{4}{3} + 1 = \frac{7}{3}.$$

The three harmonic means will be: 3, $\frac{3}{4}$, and $\frac{3}{7}$.

119. Unlimited Geometric Progressions.

The sum of a finite number of terms of a progression is always a finite number. In this section we wish to pay special attention to a geometric progression when the ratio **r** is less than 1, and the number of terms increases indefinitely.

In Sec. 116, we showed that

$$s = \frac{a(1 - r^n)}{1 - r}, \text{ where n was finite.}$$

This may be written $s = \frac{a}{1 - r} - \frac{ar^n}{1 - r}$.

Now suppose that **r** is less than 1. Then r^2 is still smaller, r^3 smaller than r^2, etc.; and if we take **n** sufficiently large, we can make r^n indefinitely small. The limit approached by r^n, as n increases without bound, is zero. Hence we define the limit of **s**, as **n** thus increases, to be $\underset{n \to \infty}{S}$, and we find that the formula becomes $\underset{n \to \infty}{S} = \frac{a}{1 - r}$. This is because $\frac{ar^n}{1 - r}$ approaches zero.

Illustration 1.

Show that the sum of infinitely many terms of the progression $1 + \frac{1}{2} + \frac{1}{4} + \ldots$, approaches the value 2.

Solution: By the formula above,

$$\underset{n \to \infty}{S} = \frac{a}{1 - r} = \frac{1}{1 - \frac{1}{2}} = \frac{1}{\frac{1}{2}} = 2.$$

This can also be seen geometrically as follows from Fig. 35.

On a coordinate system add the amounts of the terms. The first term takes us from the origin to the point 1; the sum of the first two terms is $1\frac{1}{2}$; the sum of the first three is $1\frac{3}{4}$, etc. It becomes evident that the amount which is added each time is half the remaining distance to 2. Hence by adding infinitely many of these, we shall approach two as a sum.

0 1 2

Fig. 35

Illustration 2.

Show that the repeating decimal 3.272727 . . . is the decimal result obtained from a rational fraction and find the fraction.

Solution: The given number may be written

$$3.272727\ldots = 3 + .27 + .0027 + .000027 + \ldots$$
$$= 3 + .27\,[1 + (.01) + (.01)^2 + \ldots.]$$

The quantity in brackets is a geometric progression in which a = 1 and r = .01. The sum of the series in brackets is therefore

$$s = \frac{1}{1 - .01} = \frac{1}{.99} = \frac{1}{\frac{99}{100}} = \frac{100}{99}.$$

Therefore we may now write,

$$3.272727\ldots = 3 + \frac{27}{100}\left(\frac{100}{99}\right)$$
$$= 3 + \frac{27}{99} = 3 + \frac{3}{11} = \frac{36}{11}.$$

One can easily verify the fact that the quotient obtained when 36 is divided by 11 is the repeating decimal

3.272727 . . . .

CHAPTER XVI

MATHEMATICAL INDUCTION AND BINOMIAL THEOREM

120. Introduction.

In Chapter XI we considered the topic of deductive reasoning. In this present chapter we shall consider inductive reasoning. We may define **inductive reasoning** by saying that it is a type of reasoning based upon the examination of special cases and drawing conclusions from the observations. In order that we may realize the full importance of this method, we shall consider the formula,

$$n = x^2 - x + 41. \quad (1)$$

Formula (1) is historic in mathematics. It was given at one time as a formula for computing prime numbers **n**. If one substitutes a value for x, say $x = 1$, then $n = 41$. For $x = 2$, $n = 43$. For $x = 3$, $n = 47$, etc. One may substitute many values for x and in each case compute a value of **n** which will be a prime number. If one should draw the conclusion that every value of x produces a prime number **n**, the conclusion would be erroneous. For there is at least one value of x for which **n** is not prime.

If $x = 41$, $n = 41^2 - 41 + 41 = 41^2$, which is not a prime number. We therefore examine further into this type of reasoning by studying what we shall term mathematical induction.

121. Mathematical Induction.

Proof by mathematical induction consists of two distinct steps.

Step 1. The verification of a particular or special case, say for $n = 1$.

Step 2. Showing that if the proposition holds for some particular case, say $n = k$, it is also true for the next successive case $n = k + 1$.

Both of these steps are of equal importance in the establishment of the proof.

Step 2 cannot be shown for the illustration which was given in Sec. 120.

Perhaps the matter will be clear after an illustration.

Illustration 1.

Given **n** distinct points in a plane with no three points on the same straight line, show that the number of lines necessary to join all pairs of the **n** points is given by $\frac{n(n-1)}{2}$.

Solution:

Step 1. The problem does not exist if $n = 1$. The least number of points which can be considered is $n = 2$. One line would be sufficient for joining two points. This number also satisfies the formula $\frac{n(n-1)}{2} = \frac{2(2-1)}{2} = 1$. Thus we have established the fact that the formula is true in the case $n = 2$.

Step 2. We shall now **assume** that the formula holds for some general case $n = k$. We wish to show that if it is true for $n = k$, then it holds also for $n = k + 1$. Suppose that the k points have been joined by pairs, thus using $\frac{k(k-1)}{2}$ lines. If there is a $(k + 1)$st point, all that is necessary is to join this $(k + 1)$st point to each of the **k** points, and then all points are joined by pairs. This would require **k** additional lines, so that the total number of lines would consist of the $\frac{k(k-1)}{2} + k$ lines.

$$\text{But } \frac{k(k-1)}{2} + k = \frac{k(k-1)}{2} + \frac{2k}{2}$$

$$= \frac{k(k-1+2)}{2}$$

$$= \frac{k(k+1)}{2}. \qquad (2)$$

Let us now examine the value which the formula $\frac{n(n-1)}{2}$ has when $n = k + 1$. Substituting

$n = k + 1$ we have $\frac{(k + 1)(k + 1 - 1)}{2} = \frac{(k + 1)k}{2}$, which is exactly the number found necessary from the argument which resulted in equation (2). Hence the second step has been established.

Conclusion: We have seen that the formula holds for $n = 2$ by the first step. The second step says that it holds for $n = 3$. Since it holds for $n = 3$, the second step says again that it holds for $n = 4$, etc. Consequently, we conclude that it holds for every integral value of $n > 2$.

122. Factorial Notation.

It is convenient to set up a notation for representing special types of products such as $1 \cdot 2 \cdot 3 = 6$, etc.

The symbol **k!** (sometimes written $\underline{|k}$) means **"factorial k"** and is used to indicate the product $1 \cdot 2 \cdot 3 \cdot 4 \cdot \cdot \cdot k$.

Examples:

$5!$ is read "factorial 5" and means $1 \cdot 2 \cdot 3 \cdot 4 \cdot 5 = 120$.
$(n - 1)!$ means $1 \cdot 2 \cdot 3 \cdot \cdot \cdot (n - 1)$.
$0! = 1$ (by definition).

This factorial notation will enter into the discussion of several of the remaining topics of this outline.

123. The Binomial Theorem.

By actual multiplication one may verify each of the following.

$(x + a)^2 = x^2 + 2ax + a^2$.
$(x + a)^3 = x^3 + 3ax^2 + 3a^2x + a^3$.
$(x + a)^4 = x^4 + 4ax^3 + 6a^2x^2 + 4a^3x + a^4$.
$(x + a)^5 = x^5 + 5ax^4 + 10a^2x^3 + 10a^3x^2 + 5a^4x + a^5$.

This last expression may be written in the following form:

$$(x + a)^5 = x^5 + \frac{5}{1}ax^4 + \frac{5 \cdot 4}{2!}a^2x^3 + \frac{5 \cdot 4 \cdot 3}{3!}a^3x^2 + \frac{5 \cdot 4 \cdot 3 \cdot 2}{4!}a^4x + \frac{5!a^5}{5!}.$$

This latter form gives one a clue as to the way in which $(x + a)^n$ can be written without actually performing the indicated multiplication. But before we proceed, let us note certain facts which are apparent from a study of the above cases.

1. The number of terms in the expanded form is $(n + 1)$.
2. The exponent of x in the first term of the expansion is the same as the exponent of the binomial.
3. The exponent of x decreases by unity for each successive term.
4. The first power of **a** occurs in the second term, and the exponent of **a** increases by unity for each successive term.
5. The coefficients of the terms equidistant from each end are equal.

Now it can be shown* by mathematical induction that for integral values of n,

$$(x + a)^n = x^n + nax^{n-1} + \frac{n(n-1)}{2!}a^2x^{n-2} + \frac{n(n-1)(n-2)}{3!}a^3x^{n-3}$$
$$+ \frac{n(n-1)(n-2)(n-3)}{4!}a^4x^{n-4} + \cdots$$
$$+ \frac{n(n-1)(n-2)\cdots(n-r+2)}{(r-1)!}a^{r-1}x^{n-r+1}$$
$$+ \cdots + a^n.$$

The above expansion is called the **binomial theorem.**

The term $\frac{n(n-1)(n-2)\cdots(n-r+2)}{(r-1)!}a^{r-1}x^{n-r+1}$ is called the **r^{th} term.** Any particular r^{th} term may be written down from the above expression.

Another easy way to write any r^{th} term is to make use of the following observation. Note any term of the expansion (say the fourth term) and see that in the **fourth** term every detail goes by **threes.** That is, in the fourth term, the binomial coefficient consists of 3 factors in the numerator, $n(n - 1)(n - 2)$; three factorial in the denominator; **a** has exponent 3; and **x** has exponent $n - 3$. By the same argument, every detail of the seventh term goes by sixes. Thus, if one wants the seventh term of $(x + a)^{14}$, one can write it immediately as,

* For the cases in which *n* is negative or a fraction, and for the general proof of the binomial theorem, see Fine's *College Algebra,* p. 553.

$$\frac{14 \cdot 13 \cdot 12 \cdot 11 \cdot 10 \cdot 9}{6!} a^6 x^{14-6}$$

or

$$\frac{14 \cdot 13 \cdot 12 \cdot 11 \cdot 10 \cdot 9}{6!} a^6 x^8.$$

Further observations:

a) If **n** is **even,** there are an odd number of terms in the expansion, hence a middle term.

b) If **n** is **odd,** there are an even number of terms, and there will be two central terms having the same binomial coefficient.

c) If **a** is negative, the terms involving odd powers of **a** will have a negative coefficient; *i.e.*, the expansion will consist of terms which are alternately + and −.

Illustration 1.

Expand $(2y - 3b)^4$.

Solution: Here $x = 2y$, $a = -3b$; hence by the binomial theorem,

$$(2y - 3b)^4 = (2y)^4 + 4(-3b)(2y)^3 + \frac{4 \cdot 3}{2!}(-3b)^2(2y)^2 + \frac{4 \cdot 3 \cdot 2}{3!}(-3b)^3(2y) + (-3b)^4$$

$$= 16y^4 - 96by^3 + 216b^2y^2 - 216b^3y + 81b^4.$$

Illustration 2.

Write the middle term of $(x + 3)^{12}$.

Solution: There are 13 terms in the expansion; hence the middle term is the seventh term.

But the seventh term is $\frac{12 \cdot 11 \cdot 10 \cdot 9 \cdot 8 \cdot 7}{6!} x^6 \cdot 3^{12-6} =$

$$\frac{12 \cdot 11 \cdot 10 \cdot 9 \cdot 8 \cdot 7}{6!} x^6 3^6 = \frac{12 \cdot 11 \cdot 10 \cdot 9 \cdot 8 \cdot 7}{1 \cdot 2 \cdot 3 \cdot 4 \cdot 5 \cdot 6} x^6 \cdot 729$$

$$= 673596 x^6 .$$

Illustration 3.

Find the value of $(102)^3$.

Solution: This may be written $(100 + 2)^3 =$

$$(100)^3 + 3 \cdot 2(100)^2 + 3 \cdot 2^2(100) + 2^3 =$$

$$1{,}000{,}000 + 60{,}000 + 1{,}200 + 8 = 1{,}061{,}208.$$

Illustration 4.

Use the binomial theorem to compute the first four terms of $\sqrt{1.25}$ and thereby approximate its value.

Solution:

$$\sqrt{1.25} = (1 + \tfrac{1}{4})^{1/2}$$

$$= 1^{1/2} + (\tfrac{1}{2})(1)^{-1/2}(\tfrac{1}{4}) + \frac{(\frac{1}{2})(-\frac{1}{2})}{2!}(1)^{-3/2}(\tfrac{1}{4})^2 + \frac{(\frac{1}{2})(-\frac{1}{2})(-\frac{3}{2})}{3!}(1)^{-5/2}(\tfrac{1}{4})^3 \ldots$$

$$= 1 + \tfrac{1}{8} - \tfrac{1}{128} + \tfrac{1}{1024} \ldots$$

$$= 1 + 0.125 - 0.0078125 + 0.0009765725 \ldots\ldots$$

$$= 1.1181640725 \text{ (approx.)}.$$

Four place tables should give the value as 1.1180.

If one computes the next term of the expansion (which is negative), one obtains from five terms, $\sqrt{1.25} = 1.11801 +$, and the use of additional terms does not contribute a further change in the fourth decimal place.

CHAPTER XVII

COMPLEX NUMBERS

124. Introduction.

In Chapter IX, Sec. 66, the imaginary unit $i = \sqrt{-1}$ was defined. Complex numbers, that is, numbers of the form $a + bi$ (where **a** and **b** are real), were also defined; and some of the theory of them was given for convenience in studying the quadratic equation. The reader should review Sec. 66 at this time.

125. Imaginary and Complex Numbers.

A number which is composed of a real and an imaginary part such as $a + bi$ is called a **complex number.** If **a** is zero, so that the number is of the form bi, we call such numbers **pure imaginary.**

It is also useful to note a few facts concerning powers of **i**, as displayed below.

$$\begin{aligned}
i &= \sqrt{-1} \\
i^2 &= -1 \\
i^3 &= i^2 \cdot i = -i \\
i^4 &= i^2 \cdot i^2 = (-1)(-1) = +1 \\
i^5 &= i^4 \cdot i = i \\
i^6 &= i^4 \cdot i^2 = -1 \\
i^7 &= i^4 \cdot i^3 = i^3 = -i \\
i^8 &= i^4 \cdot i^4 = +1, \text{ etc.}
\end{aligned}$$

Thus we see that successive integral powers of **i** take on a repeating sequence of values -1, $-i$, $+1$, i, etc.

The fact that every power of i which is a multiple of 4 has the value 1, enables one to make such simplifications as:

$$i^{23} = (i^4)^5 \cdot i^3 = i^3 = -i.$$

126. Fundamental Operations with Complex Numbers.

Examples of addition, subtraction, multiplication, and division will now be illustrated. Each operation will be shown for a numerical and a general case.*

Illustration 1.

Addition.

$$\text{Add:}\qquad \begin{array}{r} 2 + 4i \\ -5 + i \\ \hline -3 + 5i \end{array} \qquad\qquad \begin{array}{c} a + bi \\ c + di \\ \hline (a + c) + (b + d)i \end{array}.$$

The sum of two complex numbers is in general a complex number whose real part consists of the algebraic sum of the real parts of the two numbers, and whose imaginary part is the algebraic sum of the imaginary parts.

Illustration 2.

Subtraction.

$$\begin{array}{l} \\ \text{minus} \\ \\ \end{array}\qquad \begin{array}{r} 2 + 4i \\ -5 + 3i \\ \hline 7 + i \end{array} \qquad\qquad \begin{array}{c} a + bi \\ c + di \\ \hline (a - c) + (b - d)i \end{array}.$$

The reader should supply the statement of a rule similar to that given for addition.

Illustration 3.

Multiplication.

$$\begin{array}{rrl} 2 + & 4i & \\ -3 + & 2i & \\ \hline -6 - & 12i & \\ + & 4i & + 8i^2 \\ \hline -6 - & 8i & + 8i^2 = \\ -6 - & 8i & - 8 \;\;= \\ -14 - & 8i & \end{array} \qquad\qquad \begin{array}{rl} a + bi & \\ c + di & \\ \hline ac + bci & \\ & + adi + bdi^2 \\ \hline (ac - bd) + & (ad + bc)i \end{array}$$

Illustration 4.

Division.

The division of complex numbers involves the process of rationalization of the denominator discussed in Sec. 43, part B. The reader should review that section in connection with the following examples.

---

* For the sum and product of conjugate pairs, see Sec. 66.

$$\frac{2+4i}{3-2i} = \frac{2+4i}{3-2i}\cdot\frac{3+2i}{3+2i} = \frac{6+16i+8i^2}{9-4i^2} = \frac{-2+16i}{13}$$

$$= \frac{-2}{13} + \frac{16}{13}i.$$

$$\frac{a+bi}{c+di} = \frac{a+bi}{c+di}\cdot\frac{c-di}{c-di} = \frac{(ac+bd)+(bc-ad)i}{c^2-d^2i^2} =$$

$$\frac{(ac+bd)}{c^2+d^2} + \frac{(bc-ad)}{c^2+d^2}i.$$

All of these operations may also be performed by writing the complex numbers in trigonometric form. Since a knowledge of trigonometry has not been assumed, we shall omit this treatment of the complex numbers. Those who are interested will find the trigonometric forms of complex numbers in most texts on trigonometry and in some algebras.

127. Graphical Interpretation of Complex Numbers.

The graphical interpretation of complex numbers is accomplished by using a set of rectangular axes, such as was done with the drawing of graphs of functions. Some changes are made. The **real numbers** are laid off along the horizontal axis, called the **real axis.** The **pure imaginary** values are laid off along the vertical axis, called the **axis of imaginaries.** The unit 1 is used as a directed distance on the real axis; the unit i is used as a directed distance along the axis of imaginaries. This is done as shown in Fig. 36. The plane is called the **complex plane,** and it is to be noted that all the real numbers are now delineated by positions along the real axis. All pure imaginary numbers lie on the vertical

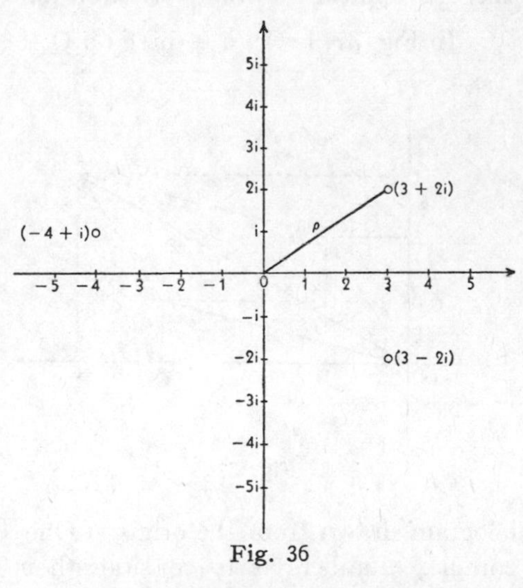

Fig. 36

axis. Complex numbers in general are then points of the plane other than those along the axes.

The figure shows the number $3 + 2i$, represented as a point 3 units over and 2 units up from the origin. If this point is joined to the origin, the length of this join, represented by ρ in the figure, is easily seen to be $\sqrt{3^2 + 2^2} = \sqrt{13}$.

In general the distance of any complex number from the origin of the complex plane is given by $\rho = \sqrt{a^2 + b^2}$ where the number is given as $a + bi$. This value $\rho = \sqrt{a^2 + b^2}$ is called the **absolute value** of the complex number, and is often written

$$\rho = |a + bi| = \sqrt{a^2 + b^2}.$$

Note that all complex numbers which lie on a circle of radius ρ about the origin will have the same absolute value, even though the values of **a** and **b** are different for each number.

Another fact to be observed is that if two complex numbers are the same, they must represent the same point. Hence if $a + bi$ and $c + di$ are equal, then $a = c$ and $b = d$.

128. Graphical Addition (or Subtraction) of Complex Numbers.

In Fig. 37, let P $(a + bi)$ and Q $(c + di)$ represent two complex numbers. Then R $(u + vi)$ represents their sum. This follows from the sum of two complex numbers as given in Sec. 126, and the fact that $OC = OA + OB$, and $OF = OD + OE$.

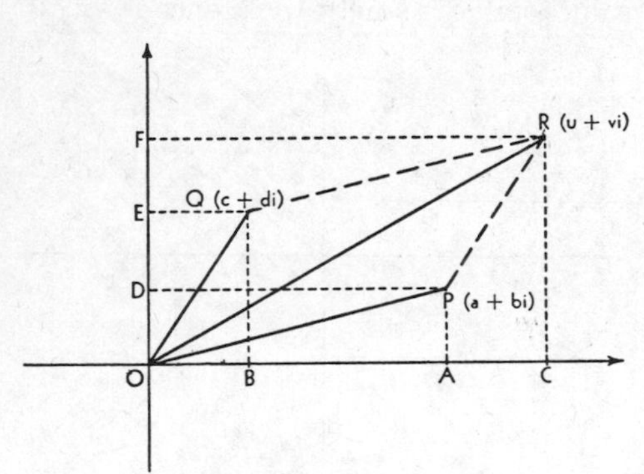

Fig. 37

The figure OPRQ is a parallelogram. The distance OR is the diagonal of this parallelogram drawn from the origin. One of the interpretations of complex numbers is to consider them as being represented by the lines such as OP, OQ, OR. This is called the **vector** representation, and the vector (such as OP) represents both magnitude and direction. Such representation is useful in problems of physics which involve forces, velocities, electric current theory, etc.

The difference of two complex numbers is easily obtained from the figure if we assume that we wish to find (u + vi) − (a + bi). This difference is given by (c + di). It is obtained graphically by adding the negative of a + bi (*i.e.*, −a − bi) to the number u + vi.

129. Some Simple Roots of Unity.*

(A) Cube roots of unity.

Consider the problem of finding the cube roots of unity. That is, we wish to find the numbers whose cube is unity. This can be stated by saying that we wish to solve the equation $x^3 = 1$.

Write this as, $x^3 - 1 = 0$.

Factor $(x - 1)(x^2 + x + 1) = 0$.

The first factor yields the solution, $x = 1$.

The second factor by the quadratic formula gives

$$x = \frac{-1 + \sqrt{-3}}{2} \quad \text{or} \quad \frac{-1 - \sqrt{-3}}{2}.$$

These are therefore the **three** cube roots of unity. Each of them has the same absolute value, namely 1. Hence they all

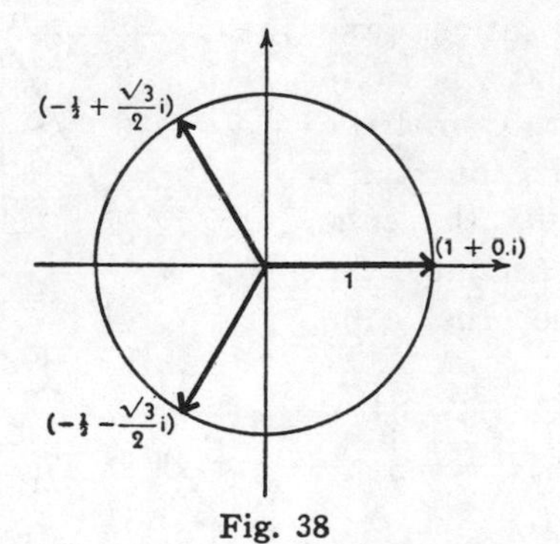

Fig. 38

lie on a circle of radius 1. Fig. 38 shows their positions. The vectors representing these three complex numbers divide the unit circle into three equal parts.

* In this connection, see Secs. 35, 41, and 136.

(B) Fourth roots of unity.

These we find by solving $x^4 - 1 = 0$.

$$x^4 - 1 = (x^2 - 1)(x^2 + 1) = (x - 1)(x + 1)(x - i)(x + i) = 0.$$

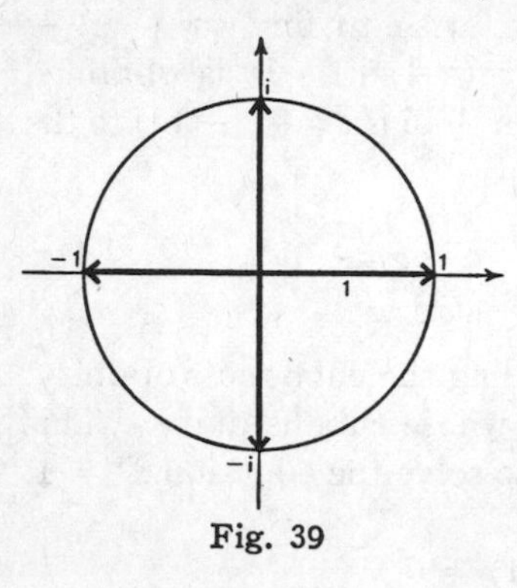

Fig. 39

The roots are $x = 1, -1, i$, or $-i$. Fig. 39 shows these values and reveals the fact that the vectors representing these solutions divide the unit circle into four equal parts.

(C) Cube roots of eight.

We solve the equation $x^3 - 8 = 0$.

$$(x^3 - 8) = (x - 2)(x^2 + 2x + 4) = 0.$$

The solutions given by these factors are $x = 2, -1 + \sqrt{3}i, -1 - \sqrt{3}i$. We see that they are each just double the cube roots of unity. Hence they each lie on a circle of radius two units and are shown in Fig. 40.

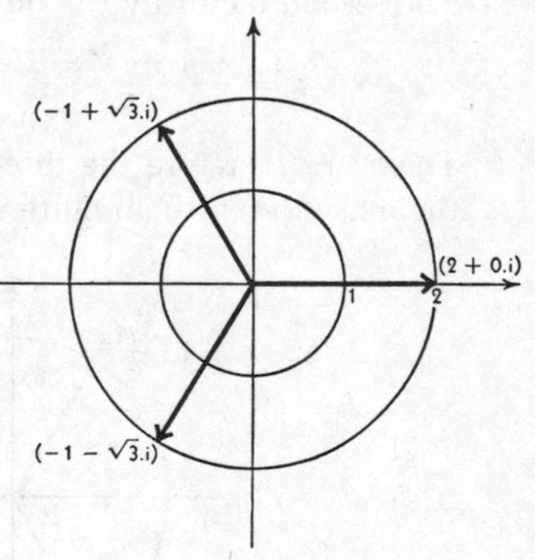

Fig. 40

These illustrations are given to show that when one has solved the problem of the cube roots of unity, one has a method for finding the cube roots of any other number.

For example, the cube roots of 17 are

$$\sqrt[3]{17}, \quad \sqrt[3]{17}\left(\frac{-1 + \sqrt{3}i}{2}\right), \text{ and } \sqrt[3]{17}\left(\frac{-1 - \sqrt{3}i}{2}\right).$$

In general the cube roots of any number A are equal to the real cube root of A, multiplied by the three cube roots of unity.

The generalizations to n^{th} roots of unity and n^{th} roots of a number A both involve considerable trigonometry and are usually not studied in a first course in algebra.

CHAPTER XVIII

THEORY OF EQUATIONS

130. Introduction.

Many students find it difficult to follow the general proofs which are usually given in college algebra for theorems in the theory of equations. For that reason this chapter will be organized mainly on the basis of numerical cases. Most of the proofs for general cases are based on the same sort of argument used in these special cases. Hence the student should be better enabled to follow them. Theorems will be stated for the general case, however, and the student will find their proofs in every college text.

There are three main reasons for studying the theory of equations.

(a) To become familiar with the theory of the polynomial, and to be able to apply it.

(b) To develop the ability to solve for the rational roots of a polynomial.

(c) To be able to approximate irrational roots of a polynomial. These three objectives are not distinct, since the last two of them depend upon theorems learned in item (a).

131. Polynomials.

A **polynomial** is an expression of the form

$f(x) = a_0x^n + a_1x^{n-1} + a_2x^{n-2} + \dots + a_{n-1}x + a_n$, where n is a positive integer and the **a's** are constants with $a_0 \neq 0$. This expression is also called a **rational integral function** of degree **n**. If the above polynomial is equated to zero, we call the resulting equation a **polynomial equation** of degree **n**.

We shall refer to

$$f(x) = a_0x^n + a_1x^{n-1} + a_2x^{n-2} + \dots + a_{n-1}x + a_n = 0 \qquad (1)$$

as a polynomial equation of degree **n**, in **first standard form.**

If a_0 is different from unity we shall find it convenient to divide both sides of equation (1) by a_0, obtaining

$$x^n + \frac{a_1}{a_0}x^{n-1} + \frac{a_2}{a_0}x^{n-2} + \ldots + \frac{a_{n-1}}{a_0}x + \frac{a_n}{a_0} = 0.$$

If we let $\frac{a_1}{a_0} = b_1, \frac{a_2}{a_0} = b_2, \ldots, \frac{a_n}{a_0} = b_n$, we have

$$x^n + b_1x^{n-1} + b_2x^{n-2} + \ldots + b_{n-1}x + b_n = 0, \quad (2)$$

which we shall call the **second standard form** of a polynomial of degree **n.**

In this chapter $f(x) = 0$ is understood to mean a polynomial of the first form unless otherwise designated. We shall consider only those cases for which the **a's** or **b's** are real numbers.

Illustration 1.

$f(x) = 3x^4 + \sqrt{5}x^3 + x^2 - \frac{2}{3}x - 14$ is a polynomial of degree 4 and the first standard form.

Illustration 2.

$f(x) = x^5 + 3x^3 - \frac{1}{4}x^2 + 6$ is a polynomial of degree 5, second standard form with $b_1 = 0$, $b_2 = 3$, $b_3 = -\frac{1}{4}$, $b_4 = 0$, $b^5 = 6$.

132. Remainder Theorem.

Consider the polynomial $f(x) = 3x^3 - 4x^2 + 6x - 8$, and suppose that this polynomial is divided by $x - 3$. We have then

$$\frac{f(x)}{x-3} = \frac{3x^3 - 4x^2 + 6x - 8}{x - 3} = 3x^2 + 5x + 21 + \frac{55}{x-3} \text{ or}$$

$$f(x) = (x - 3)(3x^2 + 5x + 21) + 55.$$

These equations express the fact that if $3x^3 - 4x^2 + 6x - 8$ is divided by $(x - 3)$, the quotient is $3x^2 + 5x + 21$ and the remainder is 55.

Let us now compute f(3) and make a comparison.

$f(3) = 3\cdot3^3 - 4\cdot3^2 + 6\cdot3 - 8 = 55 =$ Remainder [R].

Thus we have illustrated the **Remainder Theorem.**

Theorem 1. If a polynomial $f(x)$ is divided by $x - r$, the remainder is $f(r)$.

133. Factor Theorem.

Suppose one finds the value of $f(x) = x^3 - 3x^2 - 6x + 8$ when $x = 4$. That is, one finds $f(4)$.

Now $f(4) = 4^3 - 3 \cdot 4^2 - 6 \cdot 4 + 8 = 0$, and hence 4 is a root of $f(x) = 0$. This means that $f(x)$ is divisible by $x - 4$ since by the remainder theorem, $f(4) = 0$. That is, $x^3 - 3x^2 - 6x + 8 = (x - 4)(x^2 + x - 2)$. Thus we have illustrated the **Factor Theorem.**

Theorem 2. If $\mathbf{r}$ is a root of the polynomial $f(x) = 0$ (meaning that $f(r) = 0$), then $x - r$ is a factor of $f(x)$.

The converse of the factor theorem is equally true, namely:

Theorem 3. If $x - r$ is a factor of $f(x)$, then $\mathbf{r}$ is a root of $f(x) = 0$.

134. Synthetic Division.

Rather than perform the actual division, as was done in the previous examples, one finds it convenient to become familiar with a schematic device for carrying on the division. This scheme is called **synthetic division.** One writes the coefficients of $f(x)$ in order along a line. If any terms are missing in the given polynomial, one writes zero coefficients for such terms. The value of $\mathbf{r}$ is indicated to the right of this line of coefficients as shown now.

Illustration 1.

We apply this scheme to the example used in Sec. 132. The polynomial is $f(x) = 3x^3 - 4x^2 + 6x - 8$. The divisor $(x - r)$ is $x - 3$, so that $\mathbf{r}$ is 3. Thus we have the scheme,

$$\begin{array}{rrrrl} 3 & -4 & 6 & -8 & \underline{|3 = r} \\ & 9 & 15 & 63 & \\ \hline 3 & 5 & 21 & (55 = R) & \end{array}$$

These are the coefficients of the quotient.

Explanation:

Below the line one brings down the leading coefficient 3. Multiply this by the 3 in the divisor, obtaining 9, and write this 9 below the second coefficient as shown above. Find the algebraic sum, which is 5. Multiply this 5 by the divisor, 3, obtaining 15. Find the algebraic sum of 6 and 15, which is 21. Multiply 21 by 3, obtaining 63. The algebraic sum of -8 and 63 is 55, which is the remainder, and also by Theorem 1, this 55 is $f(r) = f(3)$. The other three coefficients of the third line, taken in order, are the coefficients of the quotient. It must be kept in mind that this quotient is of degree one less than the degree of the given polynomial. Hence the above scheme says that when $3x^3 - 4x^2 + 6x - 8$ is divided by $x - 3$, the quotient is $3x^2 + 5x + 21$ and the remainder is 55. It also says that $f(3) = 55$, or that the value of the function is 55 when $x = 3$.

Illustration 2.

Determine whether $x = 2$ is a root of

$$x^4 - 10x^2 + 8x + 8 = 0.$$

Solution: One must think of this polynomial equation as being of the form

$x^4 - 0 \cdot x^3 - 10x^2 + 8x + 8 = 0$, since the term in x^3 is missing.

By synthetic division,

$$\begin{array}{rrrrrl} 1 & 0 & -10 & 8 & 8 & \lfloor 2 \\ & 2 & 4 & -12 & -8 & \\ \hline 1 & 2 & -6 & -4 & (0 = R) & \end{array}$$

Since the remainder is zero, 2 is a root because $f(2) = 0$, and $x - 2$ is a factor of $f(x)$.

135. The Fundamental Theorem of Algebra.

The proof of this theorem is beyond the scope of a first course in college algebra. We state without proof*,

Theorem 4. Every polynomial $f(x) = 0$ whose coefficients are real or complex numbers has at least one root, which may be real or complex.

We shall apply this theorem only to polynomials whose coefficients are real numbers.

136. Number of Roots of a Polynomial.

The theorem of the preceding section establishes the existence of a root for every polynomial. Hence, for a given polynomial $f(x) = 0$, of degree **n**, we know that there is at least one root, say $x = r_1$. Therefore, $f(x)$ may be written,

* For a proof, see Fine's *College Algebra*, p. 588.

$$f(x) = (x - r_1)\,Q_1(x),$$

where $Q_1(x)$ is a polynomial of degree **n** − **1**. But again by the fundamental theorem, $Q_1(x)$ has a root, say $x = r_2$. Hence

$$f(x) = (x - r_1)(x - r_2)\,Q_2(x),$$

where $Q_2(x)$ is now of degree **n** − **2**. This argument can be repeated only **n** times. Then we shall have f(x) factored into **n** factors. Hence

Theorem 5. Every polynomial of degree **n** has exactly **n** roots.

Some of the roots may be repeated. Some of the roots may be complex numbers. When we say that there are **n** roots, we count each root separately even though some of them are the same numerically.

Illustration 1.

Show all the roots of the polynomial of Illustration 2, Sec. 134.

Solution: We saw that $x^4 - 10x^2 + 8x + 8 = 0$ had a root $x = 2$. Hence,

$x^4 - 10x^2 + 8x + 8 = (x - 2)(x^3 + 2x^2 - 6x - 4)$.

The remaining roots must be obtained from the quotient $x^3 + 2x^2 - 6x - 4$. Using synthetic division,

1	2	−6	−4	|2
	2	8	+4	
1	4	2	(0 = R)	

and we see that 2 is again a root. Hence,

$x^4 - 10x^2 + 8x + 8 = (x - 2)(x - 2)(x^2 + 4x + 2)$.

The remaining roots are found from the quadratic quotient $x^2 + 4x + 2$ by methods used in solving quadratic equations.

They are, $x = -2 + \sqrt{2}$ and $x = -2 - \sqrt{2}$, a pair of conjugate quadratic surds.

We have now found four roots for this fourth degree polynomial. They are,

$x = 2$, $x = 2$, $x = -2 + \sqrt{2}$ and $x = -2 - \sqrt{2}$.

We have illustrated a case for which the root 2 is repeated

and a pair of conjugate surd roots exist. That such roots enter by pairs is assured by,

Theorem 6. If a surd quantity such as $a + \sqrt{b}$ is a root of $f(x) = 0$ with real coefficients, then $a - \sqrt{b}$ is also a root. If the quadratic factors have complex roots, they also enter by pairs, as assured by,

Theorem 7. If $a + bi$ is a root of $f(x) = 0$ with real coefficients, then the conjugate value $a - bi$ is also a root.

137. Relations between Roots and Coefficients.

For use in this and future sections, we shall now establish a polynomial of degree three with roots r_1, r_2, and r_3. Such a polynomial is obtained by multiplying together the three factors obtained from the given roots. Thus

$$(x - r_1)(x - r_2)(x - r_3) = 0 \qquad \text{or}$$

$$x^3 - (r_1 + r_2 + r_3)x^2 + (r_1 r_2 + r_1 r_3 + r_2 r_3)x - r_1 r_2 r_3 = 0. \quad (3)$$

A general polynomial of degree three in second standard form may be represented as

$$x^3 + b_1x^2 + b_2x + b_3 = 0. \quad (4)$$

Let us require that equations (3) and (4) be two representations of the same polynomial. That is, we shall assume that these two polynomials have the same value for every value of x. There is a theorem which we need in this connection.

Theorem 8. If two polynomials of degree **n** are equal in value for more than **n** values of x, the two polynomials are identical. (That is, the coefficients of corresponding terms are equal.)

Hence, equating corresponding coefficients, we have

$$b_1 = -(r_1 + r_2 + r_3)$$
$$b_2 = (r_1r_2 + r_1r_3 + r_2r_3)$$
$$b_3 = -r_1r_2r_3.$$

When the coefficient of the highest degree term of a polynomial is unity, this example illustrates a theorem which may be stated as,

Theorem 9. In a polynomial of second standard form;

b_1, the coefficient of the second term, = the negative of the sum of the roots;

b_2, the coefficient of the third term, = the sum of the product of the roots taken two at a time;

b_3, the coefficient of the fourth term, = the negative of the sum of the product of roots taken three at a time;

. . . , etc.

b_n, the last term (a constant), = $(-1)^n \cdot$ (the product of all the roots).

Note: The value of $(-1)^n$ will be $-$ if **n** is odd and $+$ if **n** is even.

138. Transformation of an Equation.

Consider a third degree polynomial as given by (3), whose roots are r_1, r_2, and r_3. Suppose now we form another polynomial of degree three with roots R_1, R_2, and R_3. Similarly it may be written,

$$x^3-(R_1+R_2+R_3)x^2+(R_1R_2+R_1R_3+R_2R_3)x-R_1R_2R_3=0. \quad (5)$$

Let us further assume that the roots of equations (3) and (5) are related as shown by,

$$R_1 = 2r_1, \qquad R_2 = 2r_2, \qquad R_3 = 2r_3;$$

i.e., the roots of equation (5) are double those of equation (3). Under these circumstances equation (5) can be written

$$x^3 - (2r_1 + 2r_2 + 2r_3)x^2 + (2r_1 \cdot 2r_2 + 2r_1 \cdot 2r_3 + 2r_2 \cdot 2r_3)x - (2r_1 \cdot 2r_2 \cdot 2r_3) = 0, \qquad \text{or}$$

$$x^3 - 2(r_1 + r_2 + r_3)x^2 + 2^2(r_1r_2 + r_1r_3 + r_2r_3)x - 2^3(r_1r_2r_3) = 0. \quad (6)$$

But it will be noted that the new equation may be obtained from the given equation (3) by multiplying the second term by 2, the third term by 2^2, the fourth term by 2^3. This illustrates,

Theorem 10. To obtain a polynomial each of whose roots is equal to **k** times the corresponding roots of a given polynomial: multiply the successive coefficients beginning with the second term by $k, k^2, k^3, \ldots$, and the last term by k^n. All missing powers of x must be written into the given polynomial with zero coefficients before applying the multiplication by **k** and its powers.

Illustration 1.

Write an equation whose roots are triple the roots of $x^4 - 10x^2 + 8x + 8 = 0$.

Solution: Write the given equation as

$$x^4 + 0 \cdot x^3 - 10x^2 + 8x + 8 = 0.$$

Applying Theorem 10, we have

$$x^4 + (3)0x^3 - (9)10x^2 + (27)8x + (81)8 = 0, \quad \text{or}$$
$$x^4 - 90x^2 + 216x + 648 = 0.$$

Note: We saw in Sec. 136 that the given equation had roots 2, 2, $-2 + \sqrt{2}$, and $-2 - \sqrt{2}$. Therefore the new equation will have roots,

$$6, 6, -6 + 3\sqrt{2}, -6 - 3\sqrt{2}.$$

To obtain an equation whose roots are the **negative of the roots** of a given equation, apply Theorem 10 using $k = -1$.

Another useful transformation of an equation is one such that the new equation has roots which are equal to those of the given equation each diminished by an amount **h**.

Example: Take the given equation as $x^3 - 3x^2 - 6x + 8 = 0$, which has roots -2, 1, 4. Form a new equation such that its roots are each three less than the roots of the given equation. That is, let x_1 the new values of x which satisfy the new equation be equal to $x - 3$. One can obtain the new equation by solving $x_1 = x - 3$ for x, obtaining $x = x_1 + 3$, and replacing x in the given equation by $x_1 + 3$.

Thus, $(x_1 + 3)^3 - 3(x_1 + 3)^2 - 6(x_1 + 3) + 8 = 0$.

Simplifying and collecting terms,

$$x_1^3 + 6x_1^2 + 3x_1 - 10 = 0$$

is the required equation. Its roots are -5, -2, 1, and hence are each three less than the roots of the given equation.

The process as given involves considerable labor, but the amount of work can be shortened by using a synthetic process which is somewhat of an extension of that used in synthetic division.

This scheme is as follows. The first step consists of just those steps used heretofore. The second step and each successive one consists of synthetic division by **h** applied to the quotient obtained at each successive stage.

For example:

1	−3	−6	8	$\lfloor 3 = h$
	3	0	−18	
1	0	−6	$(-10 = R_1)$ = constant term.	
	3	9		
1	3	$(3 = R_2)$ = coef. of x_1		
	3			
1	$(6 = R_3)$ = coef. of x_1^2			
$(1$ = coef. of $x_1^3)$				

The new equation may now be written from this scheme. We obtain as before,

$$x^3 + 6x^2 + 3x - 10 = 0.$$

The method is general and can be applied to any polynomial of degree **n**, and one must remember to replace missing terms by zero coefficients.

To increase the roots by an amount **h** is equivalent to decreasing by an amount **−h** and simply means that one divides synthetically by **−h.**

Illustration 2.

Write an equation whose roots are equal to the roots of $x^4 - 10x^2 + 8x + 8 = 0$, each increased by 2.

Solution: Using the synthetic scheme:

1	0	−10	8	8	$\lfloor -2$
	−2	+4	+12	−40	
1	−2	−6	+20	$(-32 = R_1)$	
	−2	+8	−4		
1	−4	+2	$(+16 = R_2)$		
	−2	+12			
1	−6	$(+14 = R_3)$			
	−2				
1	$(-8 = R_4)$				
$(1$ = coef. of $x^4)$					

Hence the new equation is

$$x^4 - 8x^3 + 14x^2 + 16x - 32 = 0,$$

and has roots 4, 4, $\sqrt{2}$, $-\sqrt{2}$, which are each two greater than the roots of the given equation which were found in Sec. 136.

139. Possible Number of Positive and Negative Roots.

We now state a rule which is useful in determining the possible number of positive and negative roots of a polynomial. It is based upon the number of changes of signs among the terms of a polynomial. If two successive terms have different signs, one says that there is a **variation** of sign. Thus, $x^2 - 5x + 6$ has two variations of sign, from + to − and then back to +.

If $x^2 - 5x + 6$ is multiplied by $x - 1$, one obtains $x^3 - 6x^2 + 11x - 6$, which has three variations of sign. The factor $x - 1$ contributes a positive root. This last polynomial has one more variation of sign than the preceding one. Using such an argument one arrives at a rule, namely,

Descartes' rule of signs: The number of positive roots of $f(x) = 0$ cannot exceed the number of variations of sign in $f(x) = 0$. If the number of positive roots is not equal to the number of variations in sign, then the actual number is less by multiples of 2.

(This reduction is due to the fact that complex roots enter by pairs.) The number of negative roots cannot exceed the number of variations of sign in $f(-x) = 0$. The function $f(-x) = 0$ can be formed by use of Theorem 10 where $k = -1$. The positive roots of $f(-x) = 0$ are then the negative roots of $f(x) = 0$.

Illustration 1.

What information is given by Descartes' rule concerning the roots of $2x^4 + 7x^3 - 6x^2 + 8x - 3 = 0$?

Solution: By the rule for positive roots there cannot be more than 3 positive roots, since there are 3 changes of sign. Therefore, there are either 3 positive roots or 1 positive root and two complex roots.

$f(-x) = 2x^4 - 7x^3 - 6x^2 - 8x - 3$ has only one change of sign, and hence there is not more than 1 negative root. Since any variation of the number must decrease by twos, there is one negative root.

Two other consequences of Descartes' rule are:

1) If all the signs of a polynomial $f(x)$ are positive, then $f(x) = 0$ has no positive roots. (This can also be seen from another point of view. No positive value substituted for x can make a sum of terms vanish.)
2) If the signs of $f(x) = 0$ are alternately $+$ and $-$, then $f(x) = 0$ has no negative roots. (This follows because $f(-x)$ will have all signs the same.)

140. To Find the Rational Roots of a Polynomial.

Illustration 1.

Find the rational roots of

$$2x^4 + 3x^3 - 6x^2 + 8x - 3 = 0.$$

Solution: Write this equation in 2nd standard form by dividing by 2, obtaining,

$$x^4 + \tfrac{3}{2}x^3 - 3x^2 + 4x - \tfrac{3}{2} = 0. \qquad (1)$$

One may rid equation (1) of fractions by using Theorem 10 with $k = 2$. Thus,

$$x^4 + (2)\tfrac{3}{2}x^3 - (4)3x^2 + (8)4x - (16)\tfrac{3}{2} = 0 \quad \text{or}$$

$$x^4 + 3x^3 - 12x^2 + 32x - 24 = 0. \qquad (2)$$

Now, equation (2) has roots which are double the roots of the given equation. If we find these roots we can then obtain the roots of the given equation by taking half of each value found.

By Theorem 9, the product of the roots of (2) must equal -24. Hence the only possible values for roots are factors of 24, namely ±1, ±2, ±3, ±4, ±6, ±8, ±12, ±24.

One begins by using synthetic division and using the smallest values first. Using $+1$, we find

$$\begin{array}{rrrrr|l} 1 & 3 & -12 & 32 & -24 & \underline{1} \\ & 1 & 4 & -8 & +24 & \\ \hline 1 & 4 & -8 & 24 & 0 = R & \end{array}$$

so that 1 is a root and the remaining roots are solutions of the quotient $x^3 + 4x^2 - 8x + 24 = 0$.

Using $+1, -1, +2, -2, \pm 3, \pm 4, +6$ in turn we find that none of these are roots. Using -6, we have

$$\begin{array}{rrrr|l} 1 & 4 & -8 & 24 & \underline{-6} \\ & -6 & +12 & -24 & \\ \hline 1 & -2 & 4 & 0 = R & \end{array}$$

and -6 is a root. The quotient is now quadratic $x^2 - 2x + 4 = 0$ and can be solved by formula, giving $x = 1 \pm \sqrt{-3}$.

Thus we have not only found the rational roots, but we have in this instance been able to find all of the roots. They are

$$1, -6, 1 + \sqrt{-3}, \text{and } 1 - \sqrt{-3}.$$

Therefore, the roots of the original equation are

$$\tfrac{1}{2}, -3, \frac{1+\sqrt{-3}}{2}, \frac{1-\sqrt{-3}}{2}.$$

Compare the types of roots with the illustration of Descartes' rule in Sec. 139.

The methods used in this illustration may be applied to obtain all the rational roots of a polynomial. We see that the method can be made to depend upon finding **only integral roots.** Fractional values might have been used as divisors with equation (1), but this involves more work. If at any time a sufficient number of rational roots may be found to leave a quadratic quotient, then all the roots may be found since we have learned to solve the general quadratic equation. If the final quotient is of degree 5 or higher, then the problem involves methods of advanced mathematics and does not properly belong to a first course in college algebra. Moreover, the solution of a cubic or quartic polynomial in general involves methods not given here.

Illustration 2.

Find the rational roots of

$$x^4 - 5x^3 + 20x - 16 = 0.$$

Solution: According to Theorem 9, the roots must be among the factors of 16, namely among the values ± 1, ± 2, ± 4, ± 8, ± 16.

If one tries synthetic division by 1,

1	−5	0	20	−16	1
	1	−4	−4	16	
1	−4	−4	16	(0 = R)	

one finds that $x = 1$ is a root.

Division of the quotient by $x = 1$ again reveals that 1 is not a further root.

Now divide the quotient by 2,

1	−4	−4	16	2
	2	−4	−16	
1	−2	−8	(0 = R)	

and $x = 2$ is found to be a root. This last quotient $x^2 - 2x - 8$ factors into $(x - 4)(x + 2)$, and from these factors $x = -2$ or 4. The roots are all rational and are $x = +1, 2, -2, 4$.

141. The Graph of a Polynomial.

The graph can be constructed to represent a given polynomial function. This will be discussed from examples of particular cases.

Illustration 1.

Construct the graph of the polynomial

$$x^3 - 3x^2 - 6x + 8 = 0.$$

Solution:

Let $f(x) = x^3 - 3x^2 - 6x + 8$.

Now construct a chart of data as shown:

x	−3	−2	−1	0	1	2	3	4	5
f(x)	−28	0	10	8	0	−8	−10	0	28

Two methods are at our disposal in finding the values of $f(x)$ for a given value of x. One is to substitute $x = 5$, let us say, and compute $f(5) = 5^3 - 3 \cdot 5^2 - 6 \cdot 5 + 8 = 28$. The other is to use synthetic division and the remainder theorem to compute $f(r)$. Thus $f(5) = R$ or

$$\begin{array}{rrrrl} 1 & -3 & -6 & 8 & \underline{|5} \\ & 5 & 10 & 20 & \\ \hline 1 & 2 & 4 & 28 & = R = f(5) \end{array}$$

The latter method is much shorter in general, especially if the degree of the polynomial is high. The method has another advantage. Note in computing f(5) above that all the signs of the quotient and remainder are the same. If now one should compute f(6), the signs would remain positive, and f(6) would be much larger than f(5). Thus one sees that after $x = 5$, the function always increases. Furthermore, if we examine $f(-3)$ we have,

$$\begin{array}{rrrrl} 1 & -3 & -6 & 8 & \underline{|-3} \\ & -3 & +18 & -36 & \\ \hline 1 & -6 & +12 & -28 & = R = f(-3) \end{array}$$

and it appears rather obvious that $f(-4)$ will also give alternate $+$, $-$ signs but that $f(-4)$ will be negative and less

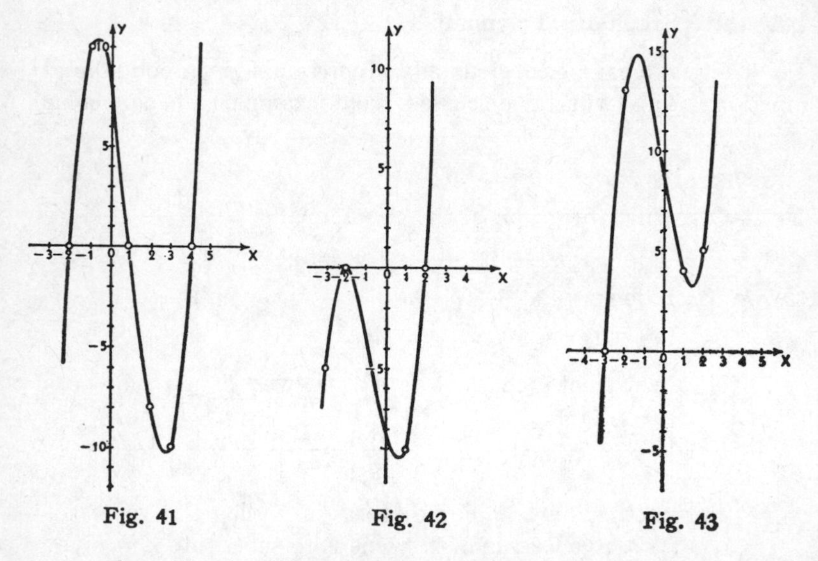

Fig. 41 Fig. 42 Fig. 43

than $f(-3)$. Hence to the left of $x = -3$, the function decreases. All the roots then lie in the interval -3 to 5. Using the data of the chart one constructs the graph of this function as shown in Fig. 41. The roots are seen to be at $x = -2, 1$, and 4.

The graph of $f(x) = x^3 + 2x^2 - 4x - 8$ (Fig. 42) shows a polynomial with roots $-2, -2, 2$. The fact that -2 is a double root accounts for the curve being tangent to the x-axis at the point $(-2, 0)$.

The graph of $f(x) = x^3 - 6x + 9$ (Fig. 43) shows a polynomial with roots -3, and $\frac{3 \pm \sqrt{3}i}{2}$. This function has only one real root, and the other two are complex. This accounts for the fact that the graph crosses the x-axis only once.

142. To Write a Polynomial with Given Roots.

Illustration 1.

Find the polynomial whose roots are $-2, 1, 4$.

Solution: The polynomial may be obtained from either of two points of view.

(A) Since the roots are -2, 1, and 4, the factors are $(x + 2)$, $(x - 1)$, and $(x - 4)$.
The product of these three factors gives
$f(x) = (x + 2)(x - 1)(x - 4) = x^3 - 3x^2 - 6x + 8$.

(B) One may use Theorem 9, as follows:
The negative of the sum of the roots is
$b_1 = -(-2 + 1 + 4) = -3$.
The sum of the products two at a time is
$b_2 = (-2\cdot1) + (-2\cdot4) + (1\cdot4) = -2 - 8 + 4 = -6$.
The product of all the roots multiplied by $(-1)^3$ is
$b_3 = -(-2\cdot1\cdot4) = +8$.
Hence we may write the polynomial,
$f(x) = x^3 + b_1x^2 + b_2x + b_3 = x^3 - 3x^2 - 6x + 8$.

Illustration 2.

Write a polynomial of lowest degree, with real coefficients, and having 1, 3, and $1 - i$ as roots.

Solution: Since $1 - i$ is a root, by Theorem 7, $1 + i$ must also be a root.

Hence the least degree must be four, and the polynomial must have roots 1, 3, $1 - i$, and $1 + i$. Therefore,

$$f(x) = (x - 1)(x - 3)(x - 1 + i)(x - 1 - i)$$
$$= (x^2 - 4x + 3)(x^2 - 2x + 2)$$
$$= x^4 - 6x^3 + 13x^2 - 14x + 6.$$

143. Approximation of Irrational Roots.

It may happen that a polynomial $f(x) = 0$ has no rational roots. An example of such a polynomial is $f(x) = x^3 - 4x^2 - x + 11$. Before demonstrating a method for approximating its roots, let us note the graph of this function. Below is a chart of values used in drawing the graph and Fig. 44 shows the graph.

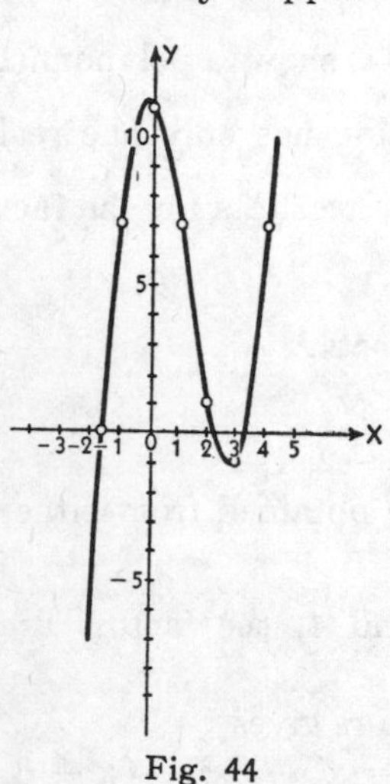

Fig. 44

x	−2	−1	0	1	2	3	4
f(x)	−11	7	11	7	1	−1	7

The method which we employ for approximating roots is based upon:

Theorem 11. If $f(x) = 0$ has a positive value for $x = \mathbf{a}$, and a negative value for $x = \mathbf{b}$ (or vice versa), then the graph of $f(x)$ has at least one root between **a** and **b**.

The interpretation may be made clear by Fig. 45, which shows $f(a)$ as positive and $f(b)$ as negative. A polynomial has a continuous graph, and therefore in order to get from P_1 to P_2 along the curve, one must cross the axis. At the point of crossing there is a root. The graph could cross more than once, but certainly must cross at least once.

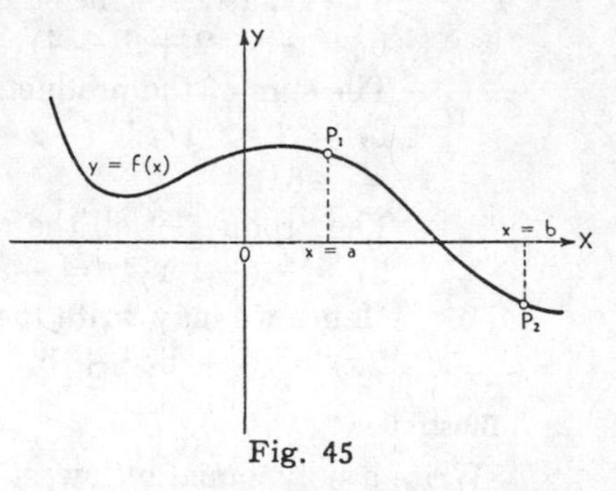

Fig. 45

The approximation is also based upon a graphical interpretation of the method given in Sec. 138 for diminishing the roots of an equation. The function used as an illustration in that section was

$$f(x) = x^3 - 3x^2 - 6x + 8 \text{ with roots } -2, 1, 4.$$

The transformed function, whose roots were each three less than those of the given function, was found to be $F(x) = x^3 + 6x^2 + 3x - 10$, and had roots $-5, -2, 1$. Fig. 46 shows the graphs

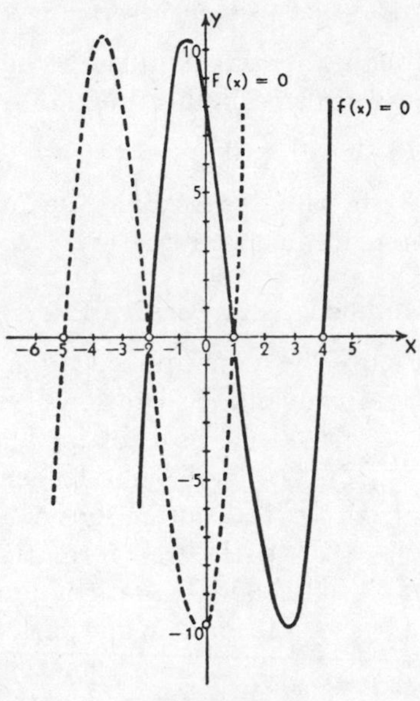

Fig. 46

of these two functions where f(x) is drawn in heavy line, the transformed function F(x) drawn in broken line.

An examination of the graph shows that the two curves are exactly the same. They differ only in their relative positions. Each point of the graph of the transformed function appears three units to the left of the corresponding point on the given function.

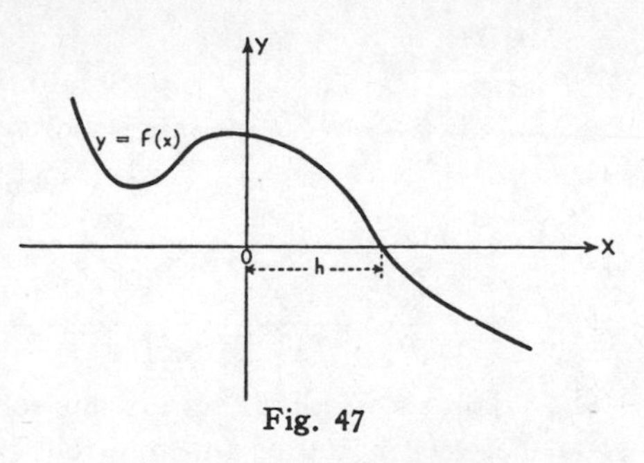

Fig. 47

The approximation depends upon a third idea. If a given function has a root at a distance **h** from the origin as shown by Fig. 47, and this function is transformed into a new one by diminishing the roots by an amount **h**, then the new function will pass through the origin.

But any function such as

$$f(x) = x^n + b_1x^{n-1} + b_2x^{n-2} + \dots + b_{n-1}x + b_n$$

which passes through the origin must be satisfied by (0, 0). Hence if (0, 0) be substituted in this function, we have,

$$0 = 0 + 0 + \dots + b_n.$$

Therefore $b_n = 0$, which means that the independent constant term of the polynomial must be zero.

144. Horner's Method. (For Positive Roots).

Horner's method is the name applied to this means of approximating irrational roots. The method will be explained in connection with a problem. It consists of successive transformations of a polynomial, where each transformation diminishes the roots by a certain amount. Attention is fixed on one particular root at a time. The transformation itself is accomplished as was shown in Sec. 138, using synthetic division.

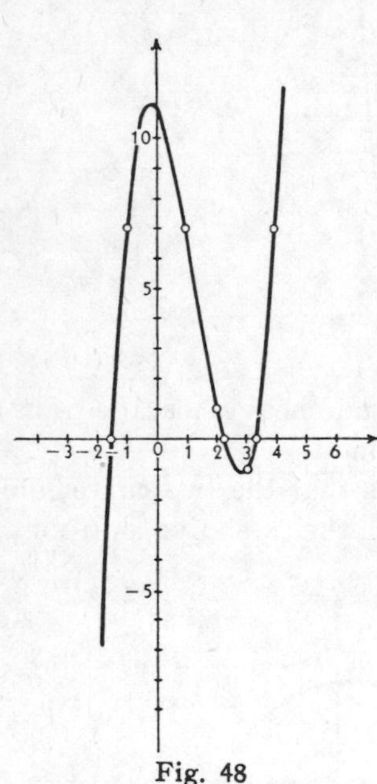

Fig. 48

Illustration 1.

Find a positive root of

$$x^3 - 4x^2 - x + 11 = 0.$$

Solution: We find by trial that this polynomial has no rational roots. We construct a chart of values and draw the graph as shown in Fig. 48.

x	−2	−1	0	1	2	3	4	5
f(x)	−11	7	11	7	1	−1	7	31

The graph shows that there is one root between −2 and −1, one root between 2 and 3, and one root between 3 and 4. Let us fix our attention on the root between 2 and 3, and form a new equation whose roots are each **two** less than the roots of the given equation.

1	−4	−1	+11	$\lfloor 2$
	2	−4	−10	
1	−2	−5	$(1 = R_1)$	
	2	0		
1	0	$(-5 = R_2)$		
	2			
1	$(2 = R_3)$			
$(1 = \text{coef. of } x^3)$				

The transformed equation whose roots are two less than those of the given equation will be written in terms of x_1 instead of x in order to indicate that this is the first transformed equation. From the synthetic scheme we see that this new equation is

$$x_1^3 + 2x_1^2 - 5x_1 + 1 = 0. \tag{1}$$

Since the given function had a root between two and three, this new function will now have a root between 0 and 1. We must now determine the new interval in which the root lies.

Finding f(.2) we have

1	2	−5	1	$\lfloor 0.2$
	0.2	0.44	−0.912	
1	2.2	−4.56	[0.088 = f(.2)]	

Finding f(.3) we have

1	2	−5	1	$\lfloor 0.3$
	0.3	0.69	−1.293	
1	2.3	−4.31	[−0.293 = f(.3)]	

and because of the change of sign of the function in passing from .2 to .3 we know by Theorem 11, that its root lies between .2 and .3.

Note: Another means of approximating the root would be to neglect all terms of equation (1) except the last two. Since x_1 is small, the terms in x_1^3 and x_1^2 would ordinarily contribute small values, and hence an approximate root could be obtained from solving $-5x_1 + 1 = 0$, yielding $x_1 = \frac{1}{5} = .2$ (approx.).

If the coefficients of the higher degree terms are large, the approximation might not be valid, in which case one would use the first method as shown for f(.2) and f(.3), etc.

Our next step is to diminish the roots of equation (1) by (0.2). Thus,

1	2	−5	+1	|0.2
	0.2	0.44	−0.912	
1	2.2	−4.56	(+0.088 = R_1)	
	.2	.48		
1	2.4	(−4.08 = R_2)		
	.2			
1	(2.6 = R_3)			
(1 = coef. of x_2^3)				

This new transformed equation is written in terms of x_2 to indicate that it is the second transformed function. It is

$$x_2^3 + 2.6x_2^2 - 4.08x_2 + 0.088 = 0. \quad (2)$$

Now a root of this equation must lie between 0 and 0.1. Since x_2 is so very small, the approximate value may be computed from the last two terms.

$$-4.08x_2 + 0.088 = 0 \text{ (approx.)} \quad \text{or}$$

$$x_2 = \frac{0.088}{4.08} = 0.02+,$$

so that a root lies between 0.02 and 0.03.

Our next step is to diminish the roots of equation (2) by 0.02. Thus,

1	2.6	−4.08	+0.088	|0.02
	.02	.0524	− .080552	
1	2.62	−4.0276	(0.007448 = R_1)	
	.02	.0528		
1	2.64	(−3.9748 = R_2)		
	.02			
1	(2.66 = R_3)			
(1 = coef. of x_3^3)				

The new equation is therefore

$$x_3^3 + 2.66x_3^2 - 3.9748x_3 + 0.007448 = 0. \qquad (3)$$

It must have a root between 0 and 0.01.

To approximate the next value with which to diminish the roots we have as before:

$$-3.9748x_3 + 0.007448 = 0 \qquad \text{or}$$

$$x_3 = \frac{0.007448}{3.9748} = 0.001+ \quad \text{(approx.)}.$$

Our next step is then to diminish the roots of (3) by 0.001. Without performing this step we may assert that the value of this root to three decimal places is

$$2 + 0.2 + 0.02 + 0.001 = 2.221$$

since this is the sum of the successive amounts by which the roots have been diminished.

Note also that the independent constant term has been approaching zero. This process may be carried on to any desired degree of approximation.

To determine the other positive root, which lies between 3 and 4, one begins by diminishing the roots the first time by 3, etc. Its value to two decimals is 3.28.

145. Negative Irrational Roots.

The problem used in Sec. 144 has a negative root, and hence we proceed to approximate it.

We pointed out in Sec. 139 that the negative roots of $f(x) = 0$ are the positive roots of $f(-x) = 0$. We use this fact now and form $f(-x)$ according to the rule given in Theorem 10, with $k = -1$.

The given function is

$$f(x) = x^3 - 4x^2 - x + 11 = 0.$$

Hence

$$f(-x) = x^3 - (-1)4x^2 - (-1^2)x + (-1^3)11 = 0$$
$$= x^3 + 4x^2 - x - 11 = 0. \qquad (4)$$

Now apply the method of Sec. 144 to find a positive root of equation (4) which lies between 1 and 2.

Diminishing the roots by 1, we have

1	4	−1	−11	1
	1	5	4	
1	5	4	(−7 = R_1)	
	1	6		
1	6	(10 = R_2)		
	1			
1	(7 = R_3)			
(1 = coef. of x_1^3)				

The first transformed equation is

$$x_1^3 + 7x_1^2 + 10x_1 - 7 = 0. \quad (5)$$

By synthetic division one discovers that there is a root of this new polynomial between 0.5 and 0.6.

Thus,

1	7	10	−7	0.5
	.5	3.75	6.875	
1	7.5	13.75	−0.125 = f(.5)	

1	7	10	−7	0.6
	.6	4.56	8.736	
1	7.6	14.56	1.736 = f(.6)	

Hence we proceed to diminish the roots of (5) by 0.5.

1	7	10	−7	0.5
	.5	3.75	6.875	
1	7.5	13.75	(−.125 = R_1)	
	.5	4.00		
1	8	(17.75 = R_2)		
	.5			
1	(8.5 = R_3)			
(1 = coef. of x_2^3)				

The new equation is now

$$x_2^3 + 8.5x_2^2 + 17.75x_2 - 0.125 = 0. \qquad (6)$$

The next approximation, obtained from the last two terms of equation (6), gives

$$x_2 = 0.007.$$

Hence one would have a root equal to $1 + 0.5 + 0.007 = 1.507$ . . . . as an approximation at this stage.

The negative root to two decimals is therefore the negative of this value, or -1.507 . . . .

Chapter XIX

PERMUTATIONS, COMBINATIONS, AND PROBABILITY

146. A Fundamental Principle.

There are three modes of transportation available in making a certain trip. These are travelling by airplane, car, or train. In how many ways can a round trip be made if one travels by a different means each way? If we let A represent airplane, C represent car, and T represent train, we may enumerate the modes of transportation for the round trip in the following ways. (The first letter represents the means of going, the second represents the means of return.)

AC, AT, CT, CA, TA, TC.

We conclude that there are 6 ways of making the round trip. If the number of cases to be examined is small, one can always analyze the problem as we have analyzed it.

However, there is another means of analysis which may be used. The going trip can be made with any one of three means of transportation, but after any one of them has been used, there remains a choice from two others for the return trip. With every choice of going there are two choices for the return trip. Hence the total number of round trips is $3 \cdot 2 = 6$. This illustrates the

Fundamental Principle. If one thing can be done in n_1 ways, and if after it has been done in any one of these ways, a second thing can be done in n_2 ways, then the two things can be done in the order stated in $n_1 \cdot n_2$ ways.

This principle may be extended to any number of cases. Thus, if one thing can be done in n_1 ways, a second in n_2 ways, a third in n_3 ways, etc., then the number of ways in which they can be done jointly in the order stated is $n_1 \cdot n_2 \cdot n_3 \ldots$.

The above principle is basic in all of the problems of this chapter.

PART 1. PERMUTATIONS

147. Definition and Formulas.

Suppose we have given **n** different objects. We choose **r** of these **n** objects and arrange the **r** chosen objects in some specified order. Each arrangement which can be formed of these r objects is called a **permutation** of the **n** objects taken **r** at a time.

We denote the number of these permutations by the symbol P(n, r), which is read, "the number of permutations of **n** things taken **r** at a time."*

Illustration 1.

An automobile dealer is offering cars made in four body styles, three types of wheels, and in five different colors. How many cars are necessary in order to display all possible schemes?

Solution: With any of the four body styles, there can be any of the three types of wheels, and any of the five different colors. Hence, he must have in stock $4 \cdot 3 \cdot 5 = 60$ cars in order to exhibit all possible cars.

By similar reasoning we may show that

$$P(n, r) = n(n-1)(n-2)(n-3)\ \ldots\ (n-r+1). \qquad (1)$$

For, the first choice can be made in any one of n ways,
the second choice can be made in any one of $n - 1$ ways,
the third choice can be made in any one of $n - 2$ ways,

. . .

. . .

. . .

. r^{th} choice can be made in any one of $n - (r - 1)$ ways.

Hence, we arrive at formula (1).

If $r = n$, we have

$$P(n, n) = n(n-1)(n-2)\ \ldots\ 2\cdot 1 = n!. \qquad (2)$$

(The symbol n! was defined in Sec. 122.)

Another form of the expression (1) can be obtained by multiplying and dividing the right-hand member by $(n - r)!$, giving

$$P(n, r) = \frac{n!}{(n-r)!}. \qquad (3)$$

* Another symbol also used is nPr.

Illustration 2.

In how many ways can six books be arranged on a shelf if two particular books must always be together?

Solution: Wrap the two particular books together to form a single package. Now the problem reduces to finding $P(5, 5)$, because we have four separate books and the package. But $P(5, 5) = 5 \cdot 4 \cdot 3 \cdot 2 \cdot 1 = 120$. For each of these 120 arrangements the two wrapped books had the same relative positions. They could be interchanged in the package and 120 other arrangements formed. The total number of arrangements is therefore $2 \cdot P(5, 5) = 2(120) = 240$.

Illustration 3.

Same as Illustration 2, except that the two particular books must never be together.

Solution: The two books are either together or separated. The total number of ways in which six books could be arranged is $P(6, 6) = 720$. We have seen above that in 240 cases the two particular books could be together. Hence, $720 - 240 = 480 =$ the number of arrangements in which the two particular books will be separated.

148. Permutations of N Things Not All Distinct.

Illustration 1.

Find the number of distinct permutations of the seven letters of the word "success."

Solution: Let P be the number of distinct permutations of the letters of the word "success." But since the two **c's** are alike, their interchange would not lead to a new permutation. Likewise the three **s's** would not lead to new permutations. In order to be able to distinguish repeated letters let us write $s_1\ u\ c_1\ c_2\ e\ s_2\ s_3$, using subscripts to aid in identification.

Because of the two **c's** and the three **s's** the number of permutations of the seven letters may be written

$$P \cdot 2! \cdot 3! = 7! \qquad \text{or} \qquad P = \frac{7!}{2! \cdot 3!}.$$

In a similar way we may show that the number of distinct permutations of **n** things taken **n** at a time, where $\mathbf{n_1}$ are alike, $\mathbf{n_2}$ are alike, etc., is

$$P = \frac{n!}{n_1!\,n_2!\,n_3!\,\dots} \tag{4}$$

149. Circular Permutations.

Illustration 1.

In how many orders can six people be arranged around a circular table?

Solution: In order to have a starting point one of the six people must be seated at some spot. Then the next seat can be filled in any one of five ways, the next in any one of four, etc. Hence the number of arrangements is $5 \cdot 4 \cdot 3 \cdot 2 \cdot 1 = 5! = 120$.

This illustrates the argument by which we may show that the number of permutations for **n** things arranged in the form of a circle is given by

$$P(n-1, n-1) = (n-1)! \tag{5}$$

PART 2. COMBINATIONS

150. Definition and Formulas.

If, from among **n** things, **r** of them are selected without regard to the order of arrangement, then any such selection is called a **combination** of the **n** things taken **r** at a time.

The symbol used is C(n, r) and is read, "the number of combinations of **n** things taken **r** at a time."

One should note the distinction between permutations and combinations. The former deals with a definite order for the **r** chosen things; the latter disregards order or arrangement. The student should exercise care in deciding whether a certain problem is one of permutation or of combination. The question of type is easily settled if one decides whether order of arrangement is essential.

Illustration 1.

Write the combinations of a, b, c, d, taken two at a time.

Solution: They may be written as ab, ac, ad, bc, bd, cd, and are six in number.

Now, ab and ba belong to the same combination, since they involve the same elements. In fact each of the combinations given above yields two permutations.

The number of permutations of four things two at a time is $P(4, 2) = 4 \cdot 3 = 12$. Hence, if one multiplies the number of combinations of four things taken two at a time by the number of permutations in each combination, the result is the number of permutations of four things taken two at a time. That is,

$$C(4, 2) \cdot P(2, 2) = P(4, 2).$$

We use this sort of argument for the general case.

Denote the number of combinations of **n** things **r** at a time by $C(n, r)$. Now these **r** things in a given combination can be arranged among themselves in $r!$ permutations. Hence,

$$r! \cdot C(n, r) = P(n, r) \qquad \text{or}$$

$$C(n, r) = \frac{P(n, r)}{r!}. \tag{6}$$

Replacing $P(n, r)$ by its value from equation (1) we have

$$C(n, r) = \frac{n(n-1)(n-2)\ \ldots\ (n-r+1)}{r!}. \tag{7}$$

If we replace $P(n, r)$ by its value from equation (3), we obtain another form,

$$C(n, r) = \frac{n!}{r!\,(n-r)!}. \tag{8}$$

Usually, we shall use equation (7) in preference to equation (8), since fewer terms need to be written.

Illustration 2.

Find **n** if $P(n, 3) = 3C(n, 4)$.

Solution: This may be written

$$n(n-1)(n-2) = \frac{3n \cdot (n-1)(n-2)(n-3)}{1 \cdot 2 \cdot 3 \cdot 4}$$

and both sides may be divided by $n(n - 1)(n - 2)$, giving

$$1 = \frac{3(n - 3)}{24} \quad \text{or}$$

$$1 = \frac{n - 3}{8} \quad \text{or}$$

$$8 = n - 3 \text{ or } n = 11.$$

The student may check the solution by writing

$$P(11, 3) = 3C(11, 4)$$

and evaluating both sides of the equation.

151. Total Number of Combinations of N Things.

Let us now write a few particular combinations.

$$C(n, 1) = \frac{n}{1} = n$$

$$C(n, 2) = \frac{n(n - 1)}{2!}$$

$$C(n, 3) = \frac{n(n - 1)(n - 2)}{3!}$$

$$C(n, 4) = \frac{n(n - 1)(n - 2)(n - 3)}{4!}$$

etc.

Comparing these values with the binomial coefficients in the binomial theorem, Sec. 123, we see that the binomial theorem for integral values of n may be written

$$(x + a)^n = x^n + C(n, 1)ax^{n-1} + C(n, 2)a^2x^{n-2} + \dots + C(n, r - 1)a^{r-1}x^{n-r+1} + \dots + C(n, n)a^n. \qquad (9)$$

Now let $x = 1$ and $a = 1$ in equation (9) and

$$(1 + 1)^n = 1^n + C(n, 1) + C(n, 2) + \dots + C(n, r - 1) + \dots + C(n, n)$$

or

$$2^n - 1 = C(n, 1) + C(n, 2) + \dots + C(n, r - 1) + \dots + C(n, n). \qquad (10)$$

Stated in words, equation 10 says that the total number of combinations of n things taken 1, 2, 3, . . . , n at a time is $2^n - 1$.

Illustration 1.

How many different sums may be composed of one each of the following: cent, nickel, dime, quarter, half-dollar, and dollar.

Solution: Since we wish to know the total number of combinations 1 at a time, 2 at a time, etc., up to 6 at a time, we have by equation (10), $2^6 - 1 = 63$. Hence 63 different sums may be composed using combinations of these six coins.

Another useful relationship among the numbers of combinations is given by

$$C(n, r) = C(n, n - r). \qquad (11)$$

This can be seen to be true from several points of view. In the first place, if we replace r by n − r in equation (8) we have

$$C(n, n - r) = \frac{n!}{(n - r)!\,r!}.$$

But this is only a rearrangement of the equation (8); hence

$$C(n, r) = C(n, n - r).$$

Secondly, if one recalls that the binomial coefficients equidistant from each end are the same, the result follows at once.

Again, one may reason that any choice of a particular combination of **r** things from **n**, is merely a discarding of **n** − **r** of the **n** things. Thus, if one makes a choice of two of the five things, a, b, c, d, e:

to choose a, b means to discard c, d, e;
to choose a, c means to discard b, d, e;
etc.,

so that the number of choices is equal to the number of discards. Hence, the conclusion expressed in equation (11).

The utility of equation (11) is shown by the following.

Illustration 2.

Compute C(15, 11).

Solution: Since C(15, 11) = C(15, 4), we would compute $C(15, 4) = \frac{15 \cdot 14 \cdot 13 \cdot 12}{1 \cdot 2 \cdot 3 \cdot 4} = 1365$, rather than

$$C(15, 11) = \frac{15\cdot14\cdot13\cdot12\cdot11\cdot10\cdot9\cdot8\cdot7\cdot6\cdot5}{1\cdot2\cdot3\cdot4\cdot5\cdot6\cdot7\cdot8\cdot9\cdot10\cdot11} = 1365.$$

Illustration 3.

In how many ways can a committee consisting of a chairman and three other members be chosen from a group of 8 men?

Solution: Suppose we choose the chairman first. He can be any one of the 8 men; hence there are 8 choices. The three other members can then be chosen from the 7 remaining men. The number of choices is $C(7, 3) = \frac{7\cdot6\cdot5}{1\cdot2\cdot3} = 35$. The number of committees is therefore $8\cdot35 = 280$.

There is a second way of solving this problem. A committee of 4 can be chosen from 8 in $C(8, 4) = \frac{8\cdot7\cdot6\cdot5}{1\cdot2\cdot3\cdot4} =$ 70 ways. In any one committee there are four choices for chairman. Hence when one indicates the chairman, there will be $4\cdot70 = 280$ committees.

PART 3. PROBABILITY

152. Definition.

A jar contains 3 black balls and 7 white balls. If we draw a ball at random from the jar, the drawing constitutes an **event.**

The events with which we shall be concerned are assumed to be **equally likely**; that is, we assume that there are no influences which would tend to force certain events.

If an event can happen in **h** different ways and fail in **f** different ways, and if all of these **h + f** ways are equally likely to occur, then the **probability of its happening** is $p = \frac{h}{h + f}$, and the **probability of its failing** is $q = \frac{f}{h + f}$.

Illustration 1.

If two balls are drawn from the jar, mentioned above, what is the probability that they will both be white?

Solution: There are $C(10, 2) = 45$ ways in which two balls can be drawn from the 10. Hence $h + f = 45$. Two white balls can be chosen from seven white ones in $C(7, 2) = 21 = h$ ways. Therefore the probability that both are white is $p = \frac{21}{45} = \frac{7}{15}$.

If **p** is the probability that an event will happen and **q** is the probability that it will fail, then

$$p + q = 1. \qquad (12)$$

For $\frac{h}{h+f} + \frac{f}{h+f} = \frac{h+f}{h+f} = 1.$

If an event is certain to happen, then $p = 1$, and $q = 0$.

We sometimes speak of the **odds** in favor of (or against) a certain event. We say that the **odds** are **p** to **q** in **favor** of an event or **q** to **p against** the event.

Illustration 2.

What are the odds against throwing a 6 with a single throw of a die?

Solution: There are 6 faces, and only one has the value of 6. The probability of throwing any other than a 6 is $\frac{5}{6}$. The probability of a 6 is $\frac{1}{6}$. Hence the odds are 5 to 1 against throwing a 6.

153. Independent Events.

If the occurrence of one event **does not** affect the probability of the occurrence of another event, the two are said to be **independent.**

Illustration 1.

A box contains 3 oranges and 4 apples. A sack contains 6 oranges and 3 apples. One object each is drawn from the box and sack. What is the probability that both are oranges?

Solution: The probability of drawing from the box does not affect the probability of drawing from the sack, so that these are independent events.

The probability p_1 of drawing an orange from the box is $\frac{3}{7}$. The probability p_2, of drawing an orange from the sack is $\frac{6}{9} = \frac{2}{3}$. The probability that both are oranges is

$$p = p_1 \cdot p_2 = \frac{3}{7} \cdot \frac{2}{3} = \frac{2}{7}.$$

By a similar argument for two independent events in which the probability of the first is p_1 and that of the second is p_2, we may prove

Theorem 1. The probability that both of two independent events will occur at a given trial is the product of their separate probabilities.

This can be extended to any number of independent events.

154. Dependent Events.

If the occurrence of one event **affects** the probability of the occurrence of a second event, the two events are said to be **dependent.**

Illustration 1.

A bag contains three apples, two oranges, and five peaches. Three objects are drawn from it in succession. What is the probability that they will be an apple, an orange, and a peach in order?

Solution: There are 10 objects in the bag. The probability of drawing an apple in one try is $\frac{3}{10}$. Suppose that an apple is drawn. The bag now contains nine objects, and the probability of drawing an orange on the next trial is $\frac{2}{9}$. Suppose an orange is drawn. Then the bag contains eight objects. The probability of drawing a peach on the next trial is $\frac{5}{8}$. Therefore the probability of drawing in order, an apple, then an orange, then a peach is by the fundamental principle

$$p = p_1 \cdot p_2 \cdot p_3 = \tfrac{3}{10} \cdot \tfrac{2}{9} \cdot \tfrac{5}{8} = \tfrac{1}{24}.$$

By generalizing this argument we may arrive at

Theorem 2. For dependent events, suppose the probability of a first event is p_1, and that after the occurrence of this event the probability of a second event is p_2, and that after the second event has occurred the probability of a third event is p_3, etc.; then the probability that all these events will occur in the prescribed order is $p = p_1 \cdot p_2 \cdot p_3 \ldots.$

155. Mutually Exclusive Events.

Two or more events are said to be **mutually exclusive** if the occurrence of one of them **excludes** the occurrence of the others.

Let us consider the case of two events, E_1 and E_2. Let e_1 and e_2 be the respective number of ways in which E_1 and E_2 can occur. The total number of ways (both favorable and unfavorable) will be represented by t. Then the probability that the event E_1 occurs is $\frac{e_1}{t}$. The probability that E_2 occurs is $\frac{e_2}{t}$. Since the events are exclusive, then the e_1 ways are all distinct from the e_2 ways. Therefore, $e_1 + e_2$ is the number of ways in which either E_1 or E_2 can occur. The probability p that one or the other of E_1 or E_2 will occur is

$$p = \frac{e_1 + e_2}{t} = \frac{e_1}{t} + \frac{e_2}{t} = p_1 + p_2.$$

Theorem 3. If **k** mutually exclusive events have the separate probabilities $p_1, p_2, \ldots\ldots p_k$, then the probability that some one of the **k** events occurs in a given trial is the sum of their probabilities.

$$p = p_1 + p_2 + \ldots + p_k.$$

Illustration 1.

In a tennis tournament, the probability that A wins is $\frac{1}{6}$ and that B wins is $\frac{2}{5}$. Find the probability that either A or B wins.

Solution: By the theorem, the probability that either A or B wins is $p = \frac{1}{6} + \frac{2}{5} = \frac{17}{30}$.

156. Repeated Trials.

Without proof*, we state two theorems.

Theorem 4. If the probability that an event will occur on a single trial is **p**, the probability that it will occur **exactly r** times among **n** trials is

$$C(n, r) \cdot p^r \cdot q^{n-r}, \text{ where } q = 1 - p.$$

Theorem 5. The probability that an event will occur at **least r** times among **n** trials is given by

$$p^n + C(n, 1)p^{n-1}q + C(n, 2)p^{n-2}q^2. \ldots\ldots + C(n, n-r)p^r q^{n-r}.$$

* For proofs, see Fine's *College Algebra*, p. 420.

Illustration 1.

What is the probability that an ace will occur **exactly** three times in five throws of a single die?

Solution: By Theorem 4,

$$p = C(5, 3)\left(\tfrac{1}{6}\right)^3 \cdot \left(\tfrac{5}{6}\right)^2 = 10 \cdot \frac{1}{216} \cdot \frac{25}{36} = \frac{125}{3888}.$$

Illustration 2.

What is the probability that an ace will occur **at least** three times in five throws of a single die?

Solution: By Theorem 5,

$$p = \left(\tfrac{1}{6}\right)^5 + C(5, 1)\left(\tfrac{1}{6}\right)^4\left(\tfrac{5}{6}\right) + C(5, 2)\left(\tfrac{1}{6}\right)^3\left(\tfrac{5}{6}\right)^2$$

$$= \left(\tfrac{1}{6}\right)^5 + 5\left(\tfrac{1}{6}\right)^4\left(\tfrac{5}{6}\right) + 10\left(\tfrac{1}{6}\right)^3\left(\tfrac{5}{6}\right)^2 = \frac{69}{1944}.$$

157. Mathematical Expectation.

If a person is to receive a sum of money M based upon the occurrence of an event whose probability of happening is **p**, then the value of his expectation E is $E = M \cdot p$.

In many instances the value of the probability **p** cannot be determined by analysis into logically existent cases, as we have done thus far. This is particularly true in applications to life and fire insurance, business forecasting, and the like. In these cases the value to be used for the probability **p** must be arrived at from statistical studies.*

For such purposes we define the following terms. If an event happens **h** times in **n** trials as shown by a statistical study, then **h** is called the **frequency of occurrence** of the event. The **relative frequency** of its occurrence is designated as $\frac{h}{n}$. If from a study of an indefinitely large number of cases, the relative frequency approaches a limit, this limit is called the **statistical** or **empirical probability**. This limit approached by $\frac{h}{n}$ is taken as the value of **p**.

Illustration 1.

A man is to receive $10 if he throws an ace with a single throw of a die. What is the value of his expectation?

* See College Outline Series—*An Outline of Statistical Methods*, by Arkin and Colton. 3rd ed. Barnes & Noble, New York.

Solution: By definition, and since the probability of throwing an ace is $\frac{1}{6}$, $E = \$10 \cdot \frac{1}{6} = \$1\frac{2}{3}$.

Illustration 2.

A man wishes to insure a house for \$6000. The insurance company knows from statistical studies that in the particular location 42 out of every 10,000 houses are destroyed annually by fire. Disregarding overhead and other factors, what premium should the man be charged for insuring his house?

Solution: This is solved on the principle of mathematical expectation. If the company insures 10,000 or more houses, it can afford to charge a premium based on the product of the insured amount and the probability of fire. The probability of fire is $\frac{42}{10000}$. Therefore the premium would amount to $\$6000 \cdot \frac{42}{10000} = \$25.20$ if other normal factors are disregarded.

Many problems of this last type arise in business. A discussion of them can be found in any text on the mathematics of finance and business.

Chapter XX

DETERMINANTS

158. Introduction.

In Chapter VIII we learned how to solve systems of linear equations. In the present chapter, we now generalize the ideas which so far have been applied to cases in which the coefficients have been definite numerical values.

Thus we begin by finding a solution for a pair of linear equations:

$$a_1x + b_1y = c_1 \tag{1}$$
$$a_2x + b_2y = c_2. \tag{2}$$

Using the method of Sec. 59: multiplying equation (1) by b_2 and equation (2) by b_1 and subtracting, we have:

$$(a_1b_2 - a_2b_1)x = c_1b_2 - c_2b_1. \tag{3}$$

In similar fashion, multiplying equation (1) by a_2 and equation (2) by a_1 and subtracting, we have

$$(a_1b_2 - a_2b_1)y = a_1c_2 - a_2c_1. \tag{4}$$

By assuming that $a_1b_2 - a_2b_1$ is different from zero, we may solve equations (3) and (4) for x and y respectively, obtaining:

$$x = \frac{c_1b_2 - c_2b_1}{a_1b_2 - a_2b_1}, \tag{5}$$

$$y = \frac{a_1c_2 - a_2c_1}{a_1b_2 - a_2b_1}. \tag{6}$$

159. Notation of Determinants.

Equations (5) and (6) involve certain typical expressions of which $a_1b_2 - a_2b_1$ is an example. A convenient notation, known as a determinant, may be used to represent this algebraic expression.

Definition 1: The symbol $\begin{vmatrix} a_1 & b_1 \\ a_2 & b_2 \end{vmatrix}$ is called a **determinant** of the **second** order, and is an identical representation for $a_1b_2 - a_2b_1$.

The four quantities involved are called the **elements** of the determinant.

Definition 2: A **determinant** is a square array of n^2 elements, arranged in **n rows** and **n columns**, the expansion of which represents an algebraic expression of **n!** terms, homogeneous in the elements of the determinant.

The rules for expanding a determinant are given later in this chapter.

A determinant of the third order, and the algebraic expression which it represents, are as follows:

$$\begin{vmatrix} a_1 & b_1 & c_1 \\ a_2 & b_2 & c_2 \\ a_3 & b_3 & c_3 \end{vmatrix} = a_1b_2c_3 - a_1b_3c_2 + b_1c_2a_3 - b_1c_3a_2 + c_1a_2b_3 - c_1a_3b_2. \quad (7)$$

160. Minors and Cofactors.

Every element of a determinant has a **minor,** which is itself a determinant of the next lower order. This minor is determined according to the following

Rule:

To find the **minor** of any particular element, strike out the row and column in which the element lies. The portion of the determinant which then remains (the elements written in their same relative positions) constitutes the minor.

Illustration 1.

Consider the determinant in equation (7) of the last section. The minor of $\mathbf{a_1}$ is $\begin{vmatrix} b_2 & c_2 \\ b_3 & c_3 \end{vmatrix}$. The minor of a_1 will be designated by the symbol A_1. So that $A_1 = \begin{vmatrix} b_2 & c_2 \\ b_3 & c_3 \end{vmatrix}$.

Illustration 2.

The minor of the element b_3 is

$$\begin{vmatrix} a_1 & c_1 \\ a_2 & c_2 \end{vmatrix} = B_3.$$

The **cofactor** of any element is defined as the minor of the element to which there has been prefixed a + or − sign, according to the

Rule:

The sign affixed is + if the sum of the row and column in which the element lies is even; the sign is − if the sum of the row and column in which the element lies is odd. Thus we multiply the minor by (-1) (to a power equal to a sum). This is + if the power of -1 is even, and − if the power is odd.

Illustration 3.

The cofactor of the element 2 in the determinant

$$\begin{vmatrix} 1 & -1 & 3 \\ 2 & 0 & -5 \\ 1 & 4 & 3 \end{vmatrix} \text{ is } (-1)^3 \begin{vmatrix} -1 & 3 \\ 4 & 3 \end{vmatrix}.$$

If one evaluates the cofactor, the result is +15, because $(-1)^3 = -1$ and the minor has the value $(-1)(3) - (3)(4) = -15$. The product of (-1) and (-15) gives $+15$.

161. The Expansion of a Determinant.

In Sec. 159, equation 7, a third order determinant was given with the algebraic expression which it represents. By means of the notion of cofactors, we are now in a position to show how the right-hand member may be obtained from the determinant.

This we proceed to formulate as follows:

Consider the elements in the first row of the determinant.
a_1 is in the first row, first column, so that the cofactor is $+A_1$.
b_1 is in the first row, second column, so that the cofactor is $-B_1$.
c_1 is in the first row, third column, so that the cofactor is $+C_1$.

$$\text{By definition then } \begin{vmatrix} a_1 & b_1 & c_1 \\ a_2 & b_2 & c_2 \\ a_3 & b_3 & c_3 \end{vmatrix} = a_1A_1 - b_1B_1 + c_1C_1 =$$

$$a_1 \begin{vmatrix} b_2 & c_2 \\ b_3 & c_3 \end{vmatrix} - b_1 \begin{vmatrix} a_2 & c_2 \\ a_3 & c_3 \end{vmatrix} + c_1 \begin{vmatrix} a_2 & b_2 \\ a_3 & b_3 \end{vmatrix} =$$

$$a_1(b_2c_3 - b_3c_2) - b_1(a_2c_3 - a_3c_2) + c_1(a_2b_3 - a_3b_2).$$

This last step reduces immediately to the right-hand member as given in (7).

Any row or column could have been used to obtain the right-hand member of (7).

Using the 2nd row, one would have,

$$-\, a_2A_2 + b_2B_2 - c_2C_2. \tag{8}$$

Using the 3rd column, one would have,

$$c_1C_1 - c_2C_2 + c_3C_3. \tag{9}$$

The reader should **verify** by actual computation that expressions (8) and (9) above give identically the right-hand member of (7). Any determinant may be expanded in terms of the elements and corresponding cofactors of any row or column.

This expansion in terms of minors is a general method. It applies also to a determinant of the second order such as $\begin{vmatrix} a_1 & b_1 \\ a_2 & b_2 \end{vmatrix}$.
For we may think of b_2 as A_1, since in this case the minor of a_1 consists of a single element, so that the expansion $a_1A_1 - a_2A_2$ becomes $a_1b_2 - a_2b_1$. Likewise for determinants of higher order:

$$\text{Thus} \begin{vmatrix} a_1 & b_1 & c_1 & d_1 \\ a_2 & b_2 & c_2 & d_2 \\ a_3 & b_3 & c_3 & d_3 \\ a_4 & b_4 & c_4 & d_4 \end{vmatrix} = a_1A_1 - a_2A_2 + a_3A_3 - a_4A_4.$$

Each minor A is of third order, and itself consists of six terms. But there are 4 such A's, and therefore the expansion of a fourth order determinant consists of $4(6) = 24$ or $4!$ terms. By extending this sort of argument a determinant of order **n** is seen to represent **n!** terms.

162. Application of Determinants.

Returning now to equations (5) and (6) we see that in terms of second order determinants we may write an equivalent form for them. Therefore the simultaneous solution of the pair of equations

$$a_1x + b_1y = c_1$$
$$a_2x + b_2y = c_2$$

can be written as

$$x = \frac{\begin{vmatrix} c_1 & b_1 \\ c_2 & b_2 \end{vmatrix}}{\begin{vmatrix} a_1 & b_1 \\ a_2 & b_2 \end{vmatrix}}, \qquad y = \frac{\begin{vmatrix} a_1 & c_1 \\ a_2 & c_2 \end{vmatrix}}{\begin{vmatrix} a_1 & b_1 \\ a_2 & b_2 \end{vmatrix}}.$$

The determinant notation gives one a systematic scheme for writing, as well as evaluating, the solution. Note carefully the pattern which is shown by the determinant solution.

Starting with three equations in three unknowns such as

$$a_1x + b_1y + c_1z = d_1 \quad (1)$$
$$a_2x + b_2y + c_2z = d_2 \quad (2)$$
$$a_3x + b_3y + c_3z = d_3 \quad (3)$$

and applying the method used in Sec. 62 for solving such a system of equations, one finally arrives at a value for x.

$$x = \frac{d_1b_2c_3 - d_1b_3c_2 + d_2b_3c_1 - d_2b_1c_3 + d_3b_1c_2 - d_3b_2c_1}{a_1b_2c_3 - a_1b_3c_2 + a_2b_3c_1 - a_2b_1c_3 + a_3b_1c_2 - a_3b_2c_1}.$$

Similarly, expressions for y and z may be obtained. But all these can be expressed in terms of third order determinants, thus,

$$x = \frac{\begin{vmatrix} d_1 & b_1 & c_1 \\ d_2 & b_2 & c_2 \\ d_3 & b_3 & c_3 \end{vmatrix}}{\begin{vmatrix} a_1 & b_1 & c_1 \\ a_2 & b_2 & c_2 \\ a_3 & b_3 & c_3 \end{vmatrix}} \qquad y = \frac{\begin{vmatrix} a_1 & d_1 & c_1 \\ a_2 & d_2 & c_2 \\ a_3 & d_3 & c_3 \end{vmatrix}}{\begin{vmatrix} \text{Same} \\ \text{as for} \\ x \end{vmatrix}} \qquad z = \frac{\begin{vmatrix} a_1 & b_1 & d_1 \\ a_2 & b_2 & d_2 \\ a_3 & b_3 & d_3 \end{vmatrix}}{\begin{vmatrix} \text{Same} \\ \text{as for} \\ x \end{vmatrix}}.$$

We see then that for systems of 2 or 3 equations the solution may be set up by determinants. Note, that when the equations are arranged so that the unknowns have the same order for each equation of the system:

a) the determinant in the **denominator** consists of the array of coefficients of the variables.

b) The determinant in the numerator always has the independent constant terms replacing the corresponding coefficients of the variables which are in the denominator. That is, if the variables are arranged in the order x, y, z, then when we are solving for x, the **d's** occur in the first column in the numerator; when we are solving for y, the **d's** are in the second column; and when we are solving for z, the **d's** are in the third column.

Illustration 1.

Find the simultaneous solution for the equations

$$3x - 4y = 7$$
$$x + 6y = 6.$$

$$x = \frac{\begin{vmatrix} 7 & -4 \\ 6 & 6 \end{vmatrix}}{\begin{vmatrix} 3 & -4 \\ 1 & 6 \end{vmatrix}} = \frac{42 - (-24)}{18 - (-4)} = \frac{66}{22} = 3.$$

$$y = \frac{\begin{vmatrix} 3 & 7 \\ 1 & 6 \end{vmatrix}}{22} = \frac{18 - 7}{22} = \frac{11}{22} = \tfrac{1}{2}.$$

Illustration 2.

Solve the system
$$x + y + z = 2$$
$$2x - y - z = 1$$
$$x + 2y - z = -3.$$

$$x = \frac{\begin{vmatrix} 2 & 1 & 1 \\ 1 & -1 & -1 \\ -3 & 2 & -1 \end{vmatrix}}{\begin{vmatrix} 1 & 1 & 1 \\ 2 & -1 & -1 \\ 1 & 2 & -1 \end{vmatrix}} = \frac{2\begin{vmatrix} -1 & -1 \\ 2 & -1 \end{vmatrix} - 1\begin{vmatrix} 1 & 1 \\ 2 & -1 \end{vmatrix} + (-3)\begin{vmatrix} 1 & 1 \\ -1 & -1 \end{vmatrix}}{1\begin{vmatrix} -1 & -1 \\ 2 & -1 \end{vmatrix} - 2\begin{vmatrix} 1 & 1 \\ 2 & -1 \end{vmatrix} + 1\begin{vmatrix} 1 & 1 \\ -1 & -1 \end{vmatrix}} =$$

$$\frac{2(1+2) - 1(-1-2) - 3(-1+1)}{1(1+2) - 2(-1-2) + 1(-1+1)} = \frac{9}{9} = 1.$$

[Both determinants have been expanded in terms of elements of the first column.]

The solutions for y and z are set up and have values as follows:

$$y = \frac{\begin{vmatrix} 1 & 2 & 1 \\ 2 & 1 & -1 \\ 1 & -3 & -1 \end{vmatrix}}{9} = \frac{-9}{9} = -1, \qquad z = \frac{\begin{vmatrix} 1 & 1 & 2 \\ 2 & -1 & 1 \\ 1 & 2 & -3 \end{vmatrix}}{9} = \frac{18}{9} = 2.$$

The above methods are applicable also to any system consisting of **n** equations in **n** unknowns.

163. Properties of Determinants.

For convenience we shall designate a typical third order determinant by D. The properties of determinants as stated below hold for **any determinant of order n.** However, general proof will not be given, but the reader will be shown how to verify the theorems for third order determinants.

$$\text{Let } D = \begin{vmatrix} a_1 & b_1 & c_1 \\ a_2 & b_2 & c_2 \\ a_3 & b_3 & c_3 \end{vmatrix}. \quad \text{Then,}$$

Theorem 1. If rows are made into columns in the same orders, the value of a determinant is unchanged.

That is, $\begin{vmatrix} a_1 & b_1 & c_1 \\ a_2 & b_2 & c_2 \\ a_3 & b_3 & c_3 \end{vmatrix} = \begin{vmatrix} a_1 & a_2 & a_3 \\ b_1 & b_2 & b_3 \\ c_1 & c_2 & c_3 \end{vmatrix}$, and the reader should verify this by actually expanding both determinants.

This theorem gives one the right to substitute the word **column for row** in any theorem on determinants.

Theorem 2. If two columns (or rows) of a determinant are the same, the value of the determinant is zero.

The reader should evaluate $\begin{vmatrix} a_1 & b_1 & c_1 \\ a_1 & b_1 & c_1 \\ a_3 & b_3 & c_3 \end{vmatrix}$ to see this fact.

Theorem 3. Any factor **k** may be removed from all the elements of a column (or row), if the residual determinant is multiplied by the value **k.**

That is, $\begin{vmatrix} ka_1 & b_1 & c_1 \\ ka_2 & b_2 & c_2 \\ ka_3 & b_3 & c_3 \end{vmatrix} = k \begin{vmatrix} a_1 & b_1 & c_1 \\ a_2 & b_2 & c_2 \\ a_3 & b_3 & c_3 \end{vmatrix}$,

as the reader may verify by actually expanding both determinants.

Theorem 4. If all the elements of a column (or row) of a determinant are multiplied by the same number **k,** the value of the determinant is multiplied by **k.**

(This theorem follows from the demonstration suggested in Theorem 3.)

Theorem 5. The value of a determinant is unchanged if each element of any column (or row) is multiplied by a value **m**

and added to or subtracted from the corresponding elements of another column (or row).

To see this, let the reader verify the following

$$\begin{vmatrix} (a_1 + mb_1) & b_1 & c_1 \\ (a_2 + mb_2) & b_2 & c_2 \\ (a_3 + mb_3) & b_3 & c_3 \end{vmatrix} = \begin{vmatrix} a_1 & b_1 & c_1 \\ a_2 & b_2 & c_2 \\ a_3 & b_3 & c_3 \end{vmatrix} + m\begin{vmatrix} b_1 & b_1 & c_1 \\ b_2 & b_2 & c_2 \\ b_3 & b_3 & c_3 \end{vmatrix} = \begin{vmatrix} a_1 & b_1 & c_1 \\ a_2 & b_2 & c_2 \\ a_3 & b_3 & c_3 \end{vmatrix}.$$

Note that the determinant (above) which is multiplied by **m**, has the value zero because of Theorem 2.

The verification for subtracting will involve writing $(a_1 - mb_1)$, etc., and is quite similar.

164. Application of the Theorems.

Illustration 1.

$$\text{Evaluate} \quad \begin{vmatrix} 2 & 5 & 1 \\ 3 & 4 & 1 \\ 6 & 1 & 1 \end{vmatrix}.$$

Using Theorem 5, add columns 1 and 2 to form a new determinant,

$$\text{namely,} \quad \begin{vmatrix} 7 & 5 & 1 \\ 7 & 4 & 1 \\ 7 & 1 & 1 \end{vmatrix}.$$

Now divide the first column by 7, by Theorem 3, obtaining

$$7 \begin{vmatrix} 1 & 5 & 1 \\ 1 & 4 & 1 \\ 1 & 1 & 1 \end{vmatrix}, \text{ which is equal to } 7(0) \text{ because}$$

by Theorem 2, this last determinant has the value zero, since the 1st and 3rd columns are alike. Hence the value of the given determinant is zero.

Illustration 2.

$$\text{Evaluate} \quad \begin{vmatrix} 2 & -3 & 1 \\ -1 & 2 & 1 \\ 3 & 1 & 2 \end{vmatrix}.$$

Write a new determinant from the given one by the following steps, each of which will not change the value of the determinant.

1. Copy first row as given.
2. Form a new 2nd row by subtracting row 1 from row 2. (Theorem 5)
3. Form a new 3rd row by multiplying row 1 by -2 and add to row 3. (Theorem 5)

Thus we obtain $\begin{vmatrix} 2 & -3 & 1 \\ -3 & 5 & 0 \\ -1 & 7 & 0 \end{vmatrix}$.

Expand this new determinant in terms of the elements of the 3rd column.

Thus $\begin{vmatrix} 2 & -3 & 1 \\ -3 & 5 & 0 \\ -1 & 7 & 0 \end{vmatrix} = 1 \begin{vmatrix} -3 & 5 \\ -1 & 7 \end{vmatrix} - 0 \cdot \begin{vmatrix} \text{minor} \end{vmatrix} + 0 \cdot \begin{vmatrix} \text{minor} \end{vmatrix}$

$$= \begin{vmatrix} -3 & 5 \\ -1 & 7 \end{vmatrix} = -3(7) - 5(-1) = -21 + 5 = -16$$

This illustration shows a third order determinant represented by an equivalent second order determinant. This representation was brought about by using the theorems to **insert** zeros conveniently so that when an expansion was made in terms of the elements of some row or column, all except one of the minors would be multiplied by zero. By utilizing this idea any determinant of order **n** can be represented by an equivalent 2nd order determinant. Practice and considerable insight are somewhat essential in such procedure.

Illustration 3.

Evaluate $\begin{vmatrix} 2 & 0 & 3 & 3 \\ 0 & -1 & 2 & 1 \\ 1 & -1 & 3 & 1 \\ 0 & 3 & 1 & 2 \end{vmatrix}$.

Without using any of the theorems, this determinant can be expanded quite conveniently in terms of the elements of the first column, since this column contains two zeros which will simplify the expansion.

Thus we would have, $2 \begin{vmatrix} -1 & 2 & 1 \\ -1 & 3 & 1 \\ 3 & 1 & 2 \end{vmatrix} + 1 \begin{vmatrix} 0 & 3 & 3 \\ -1 & 2 & 1 \\ 3 & 1 & 2 \end{vmatrix} = -16.$

The reader should verify this result by actual expansion. A second approach would be to copy the last three rows as given. Form a new first row by multiplying the 3rd row by -2 and adding to row 1. Thus we obtain,

$$\begin{vmatrix} 0 & 2 & -3 & 1 \\ 0 & -1 & 2 & 1 \\ 1 & -1 & 3 & 1 \\ 0 & 3 & 1 & 2 \end{vmatrix}.$$

Expanding now in terms of elements of the first column, the only minor left is

$$1\begin{vmatrix} 2 & -3 & 1 \\ -1 & 2 & 1 \\ 3 & 1 & 2 \end{vmatrix}.$$

But this is just the original determinant given in Illustration 2, and has the value -16.

165. Linear Systems of Homogeneous Equations.

Definition: If all the terms are of the same degree in the unknowns, an equation is said to be **homogeneous.**

Consider the homogeneous system,

$$\begin{aligned} a_1x + b_1y + c_1z &= 0 \\ a_2x + b_2y + c_2z &= 0 \\ a_3x + b_3y + c_3z &= 0. \end{aligned}$$

These are rather obviously all satisfied by the solution $x = 0$, $y = 0$, $z = 0$, called the **trivial solution.** If there is any other solution then some minor, say $\begin{vmatrix} a_1 & b_1 \\ a_2 & b_2 \end{vmatrix}$, must not vanish when

$$\begin{vmatrix} a_1 & b_1 & c_1 \\ a_2 & b_2 & c_2 \\ a_3 & b_3 & c_3 \end{vmatrix} = 0.$$

In this case, then, the first two equations may be solved for x and y in terms of z, giving

$$x = \frac{\begin{vmatrix} -c_1z & b_1 \\ -c_2z & b_2 \end{vmatrix}}{\begin{vmatrix} a_1 & b_1 \\ a_2 & b_2 \end{vmatrix}}, \qquad y = \frac{\begin{vmatrix} a_1 & -c_1z \\ a_2 & -c_2z \end{vmatrix}}{\begin{vmatrix} a_1 & b_1 \\ a_2 & b_2 \end{vmatrix}}.$$

If then a solution other than the trivial one exists, there will be an infinite number of solutions. Since x and y are each expressed in terms of z, then when any value different from zero is assigned to z, a pair of values of x and y are determined.

166. Systems in N Equations in (N-1) Unknowns.

Theorem. A **necessary** condition that **n** equations in **n-1** unknowns have a solution is that the eliminant of the system vanish.

Definition: The determinant which consists of the array of coefficients and the constant terms is called the **eliminant** of the system.

We shall not prove this theorem for the case of **n** equations, but we shall illustrate the general case by a proof for three equations in two unknowns.

Consider the three equations

$$a_1x + b_1y = c_1$$
$$a_2x + b_2y = c_2$$
$$a_3x + b_3y = c_3.$$

Recalling that these equations each represent lines, we should not ordinarily expect a solution of the first two equations to satisfy the third equation. In order for this to happen, the three lines would have to intersect in a common point.

Suppose, however, that they have a common solution. Solving the first two equations, for x and y, yields,

$$x = \frac{\begin{vmatrix} c_1 & b_1 \\ c_2 & b_2 \end{vmatrix}}{\begin{vmatrix} a_1 & b_1 \\ a_2 & b_2 \end{vmatrix}}, \qquad y = \frac{\begin{vmatrix} a_1 & c_1 \\ a_2 & c_2 \end{vmatrix}}{\begin{vmatrix} a_1 & b_1 \\ a_2 & b_2 \end{vmatrix}}.$$

Since these values must, by assumption, satisfy the third equation, we have

$$a_3\frac{\begin{vmatrix} c_1 & b_1 \\ c_2 & b_2 \end{vmatrix}}{\begin{vmatrix} a_1 & b_1 \\ a_2 & b_2 \end{vmatrix}} + b_3\frac{\begin{vmatrix} a_1 & c_1 \\ a_2 & c_2 \end{vmatrix}}{\begin{vmatrix} a_1 & b_1 \\ a_2 & b_2 \end{vmatrix}} = c_3$$ or, clearing fractions,

$$a_3\begin{vmatrix} c_1 & b_1 \\ c_2 & b_2 \end{vmatrix} + b_3\begin{vmatrix} a_1 & c_1 \\ a_2 & c_2 \end{vmatrix} - c_3\begin{vmatrix} a_1 & b_1 \\ a_2 & b_2 \end{vmatrix} = 0,$$ which is just an expansion

of the determinant $\begin{vmatrix} a_1 & b_1 & c_1 \\ a_2 & b_2 & c_2 \\ a_3 & b_3 & c_3 \end{vmatrix} = 0$, thus proving the theorem for this case. We say that such a system is **consistent**.

Illustration.

Determine whether the system

$$\begin{aligned} 3x - 4y &= 7 \\ x + 6y &= 6 \\ x - 2y &= 2 \end{aligned}$$

has a solution.

Forming the eliminant $\begin{vmatrix} 3 & -4 & 7 \\ 1 & 6 & 6 \\ 1 & -2 & 2 \end{vmatrix}$, the student can readily verify that its value is zero. This is a necessary condition but not a sufficient condition. However, we may solve the first two of the given equations simultaneously, and if the values found satisfy the third equation we shall know that the system has a solution.

Solving $3x - 4y = 7$

$x + 6y = 6,$

we find $x = 3$, $y = \frac{1}{2}$.

These equations we solved in Sec.162 of this chapter.

Substituting these values of x and y in the third equation, yields,

$$3 - 2(\tfrac{1}{2}) = 2,$$

and hence the system has a solution. The geometric interpretation is that the three lines are concurrent.

APPENDIX

PART A

TO PROVE THAT $\sqrt{2}$ IS AN IRRATIONAL NUMBER

In Sec. 39 we defined rational and irrational numbers. We wish now to establish the fact that $\sqrt{2}$ is an irrational number.

Proof: Let us assume that $\sqrt{2}$ is rational, so that

$$\sqrt{2} = p/q \qquad (1)$$

where p and q are two integers which are relatively prime.

Square equation (1), obtaining,

$$2 = p^2/q^2 \text{ or}$$
$$2q^2 = p^2. \qquad (2)$$

Equation (2) tells us that p^2 is an even integer, since it is twice another integer. If p^2 is an even number, then p is an even number.

[The even integers are 0, 2, 4, 6, 8, or any integer which ends in these integers. The square of any even integer is even. Odd integers are 1, 3, 5, 7, 9, or any integer ending in these integers. The square of an odd integer is odd.]

Since p is even, it has a factor 2, and p^2 must have a factor 4.

Now divide both sides of equation (2) by 4, obtaining

$$2q^2/4 = p^2/4 = k \text{ or}$$
$$q^2/2 = k \qquad \text{or}$$
$$q^2 = 2k. \qquad (3)$$

[We call $p^2/4 = k$, because we know that p^2 is divisible by 4.]

From equation (3) it follows by a similar argument that q^2 and q are even integers. Hence p and q are not relatively prime. Therefore, we are led to a contradiction of our assumption, and $\sqrt{2}$ is not a rational number.

Hence $\sqrt{2}$ is irrational.

The method of argument used in this proof is called **reductio ad absurdum.**

Part B

THE NUMBER SYSTEM OF ALGEBRA

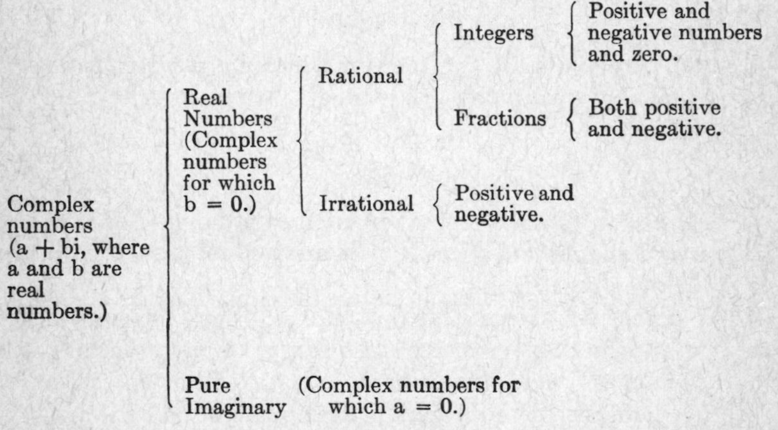

This chart shows the relationships among the kinds of numbers we have used. Every number may be thought of as complex, yet we usually write some numbers according to the following agreements.

1. i represents $\sqrt{-1}$.
2. $\frac{2}{3} + 0 \cdot i$ we write as $\frac{2}{3}$.
3. $0 + bi$ we write as bi.
4. $0 + 0 \cdot i$ we write as 0.
5. $1 + 0 \cdot i$ we write as 1.
6. $0 + 1 \cdot i$ we write as i.

FINAL EXAMINATIONS

The next few pages contain copies of eight final examinations. They are included for the purpose of aiding the student in preparing himself for examination.

The answers to the examinations are given on pages 219 ff. Whenever the answers involve definitions or proofs, reference is given to pages in the Outline.

Suggestions for the Use of the Questions.

In preparing for an hour quiz on special topics, pick the questions dealing with any particular subject from the first six examinations. Solve the questions and check the results. Use the outline and its illustrations to clarify obscure matters.

When preparing for a final examination, use the first six sets of questions, solving each for practice and review. Then find time (about three hours) to sit down with Examination G and work the problems as you would on an examination. Check your work by using the answers given. If you have difficulty with some questions, study the outline on those parts again. Repeat this procedure with Examination H, and you should be well prepared for your final examination.

EXAMINATION A

Answer any ten.

1. Define function.

   Solve for x: $\sqrt{x+1} = \sqrt{9-x} - \sqrt{x+4}$.

2. Develop the formula for solving a quadratic equation.

   Define and discuss the discriminant of a quadratic equation.

3. Simplify:

(*a*) $\dfrac{X^2 - 1/X}{X + 1/X + 1}$ (*b*) $\dfrac{1+3i}{1-5i}$ (*c*) $(A^{1/3} - A^{-1/3})(A^{2/3} + A^{-2/3} + 1)$.

4. Solve simultaneously and check by graphing:

$$x^2 + y^2 = 25 \qquad x + y = 5.$$

5. Given the equation $2X^2 + 3kX + k + 1 = 0$. Find "k" so that
   (*a*) the equation has equal roots.
   (*b*) the sum of the roots is six.
   (*c*) one root shall be zero.
   (*d*) the product of the roots is nine.

6. (*a*) Find $\log \sqrt[3]{\frac{(1.5)\,(60)^2}{(2.1)}}$ (*b*) Solve for x if $3^x = 7$.

7. Write the first five and the eleventh terms of $(2x - y)^{21}$. Define logarithm.

8. Evaluate: $\begin{vmatrix} 2 & 1 & -1 & 1 \\ 1 & 2 & 3 & 1 \\ 2 & -1 & 1 & 2 \\ 1 & 1 & 1 & 1 \end{vmatrix}.$

9. Define complex number. State two theorems concerning complex numbers. Show by substitution that $\frac{-1 - \sqrt{3}\,i}{2}$ is a root of $x^3 - 1 = 0$.

10. State: Descartes' rule of signs; the remainder theorem; the fundamental theorem of algebra; the factor theorem. Prove one of them.

11. (*A*) State as a theorem the relation between the roots of a polynomial and its coefficients.
    (*B*) In how many ways can a set of 10 numbered sections be chosen from this examination?

12. Solve for all of the roots of:

$$X^5 + 3X^4 + 2X^3 - X^2 - 3X - 2 = 0.$$

13. Solve for "y" only, by determinants:

$$\begin{aligned} x + y + 2z &= 0 \\ 2x - y - 2z &= -1 \\ 3x - 5y + 6z &= -1. \end{aligned}$$

EXAMINATION B

Answer the first six questions. Omit two of the last six questions.

1. Solve the following equations:

   (*a*) $\sqrt{x+1}+\sqrt{x-1}=2$ (*b*) $\frac{1}{2}X^{-2/3}=2.$

2. *A*. Given $f(x)=\dfrac{x^3-2x^2+1}{x-1}$. Find (*a*) f(3); (*b*) f(−2).

   B. The base of a triangle is one unit more than the altitude. Express the area of the triangle as a function of the altitude.

3. *A*. Solve simultaneously: $x^2+y^2=9$
   $x+y=1.$

   B. Draw the graphs of the two equations in (*A*).

4. *A*. Find (*a*) log 12
   (*b*) $\log\sqrt{1.8}$
   (*c*) $\log\dfrac{.003}{2}$
   (*d*) $\log 6\ (3000)^5$.

   B. Solve for x: $12^x=18$.

5. Find all the roots of the equation: $2x^3+3x^2+3x+1=0$

6. State and prove the remainder theorem.

7. *A*. Derive the formula for the sum of the first n terms in a geometrical progression.

   B. Find the sum of the infinite series:

$$\frac{1}{2}-\frac{1}{4}+\frac{1}{8}-\frac{1}{16}+\frac{1}{32}-\cdots$$

8. The plate of a mirror is 18 inches by 12 inches, and it is to be framed with a frame of uniform width whose area is to be equal to that of the glass. Find the width of the frame.

9. By means of determinants solve for y:

$$\begin{aligned} 2x-3y-4z &= 1\\ x+2y-5z &= -3\\ -x-2y+4z &= 3. \end{aligned}$$

10. *A.* Write the first 5 terms, the r-th term, and the last term of the expansion of $(x + y)^n$.

    B. Expand: $(2x - 1)^5$. Simplify each term.

11. *A.* If z varies inversely as the square of y and if $z = 4$ when $y = 3$, find z when $y = 6$.

    B. Determine k so that $x^2 + kx + 4 = 0$ shall have: (*a*) equal roots (*b*) one root equal to 3.

12. *A.* How many distinct permutations can be made of the letters of "football"?

    B. A box contains 5 black and 3 white balls. Two balls are drawn out at random. What is the probability (*a*) that both balls are black? (*b*) that one ball is black? (*c*) that both balls are white?

EXAMINATION C

Answer ten questions.

1. (*a*) Define function.
   (*b*) Graph $f(x) = x^3 - 7x + 6$.
2. (*a*) Solve $ax^2 + bx + c = 0$ by completing the square.
   (*b*) Determine k in the equation $kx^2 + 4x + 1 = 0$ so that
      (1) the roots will be real and equal,
      (2) the sum of the roots will be 2.
3. Solve for x and y, $x^2 + y^2 - 11x + y - 2 = 0$
   $x - y + 2 = 0$.
4. (*a*) Determine the equation whose roots are $(1 + 2i)$, $(1 - 2i)$, and 3.
   (*b*) Find the roots of $\sqrt{x} = \sqrt{x + 8} - \sqrt{x + 3}$.
5. (*a*) Expand $\left(\frac{1}{\sqrt{3}} - \sqrt{3}\right)^6$.
   (*b*) Find the tenth term of the expansion of $(x + \sqrt{y})^{19}$.
6. (*a*) Find $\log_{10}\sqrt{14.7}$.
   (*b*) Find $\log_3 7$.
   (*c*) Prove $\log_a uv = \log_a u + \log_a v$.

7. (*a*) State and prove the remainder theorem.
   (*b*) Given the equation $x^4 - 3x^3 - 2x^2 + 6x + 4 = 0$, apply Descartes' rule of signs and find all the roots.

8. (*a*) Insert four arithmetic means between 2 and 12.
   (*b*) In a geometric progression: $a = 2$
   Find l and s. $r = 3$
   $n = 6$

9. Find the probability of drawing 2 white and 4 black balls in a single draw from a bag containing 6 white and 7 black balls.

10. (*a*) Set up the determinants for finding the value of x in the following equations:

$$2x - y + 2z = 5$$
$$y - 2z - 3w = -7$$
$$x + y + z + 3w = 3$$
$$x - 2y + 7w - 7 = 0.$$

   (*b*) Evaluate the denominator.

11. If a heavier weight draws up a lighter one along a rough inclined plane by means of a string passed over a fixed wheel, the space described in a given time varies directly as the difference between the weights and inversely as their sum. If 9 ounces draws 7 ounces through 8 feet in 2 seconds, how far up will 12 ounces draw 9 ounces in the same time?

12. (*a*) Add graphically $(5 + 2i)$ and $(-1 + 7i)$.
   (*b*) Express as a complex number in the form $a + bi$,

$$\frac{4 + 3i}{2 - i}.$$

EXAMINATION D

Answer any ten questions.

1. Solve the equation $x^2 + 6x + 34 = 0$.
   Check by substituting the values found in the equation.
   Reduce $\frac{3 - 2i}{4 + 3i}$ to a complex number of the form $a + bi$.

2. Solve $f(x) = x^2 - x - 6 = 0$ by "completing the square." Check by graphing the function.
   Solve $\sqrt{2x + 6} = 3x - 11$.

3. Solve simultaneously:
   (*a*) $x^2 + y^2 = 10$, $y - 2x + 5 = 0$ (*b*) $x^2 + 2xy + y^2 = 28$, $xy = 8$.

4. Find $\log \sqrt[3]{\dfrac{(50)^2\ (.6)\ 9}{21}}$.
   Define logarithm.

5. The intensity of illumination varies inversely as the square of the distance from the source of light. If the intensity at a distance of 12 feet is 30 candle power, what is it at 20 feet?

6. Define arithmetic, geometric, and harmonic progressions; and give an example of each.
   Find the sum of the even numbers between 1 and 100.

7. Derive two of the following:
   Formula for sum of geometric progression.
   Formula for sum of arithmetic progression.
   Formula for solving a quadratic equation.

8. Write the first four terms of $(2x - 3)^{51}$.
   State three theorems applying to determinants.

9. Solve for "z" only, by determinants:

$$\begin{aligned} 3x + 4y - 2z &= 5 \\ 4x - 3y + 8z &= -4 \\ 2x + 8y - 3z &= 5. \end{aligned}$$

10. Find all of the roots of $x^5 + 3x^4 + 2x^3 - x^2 - 3x - 2 = 0$.
    Write an equation whose roots are the roots of the above equation each diminished by 2.
    Write an equation whose roots are double the roots of the given equation.

11. In how many ways can three prizes be awarded in a bridge contest in which 20 participate, if only one prize goes to an individual?

In how many ways can a committee of 5 be chosen from 6 men and 5 women, if three men and two women are to be on each committee?

12. State Descartes' rule of signs. State the fundamental theorem of algebra. State the remainder theorem, and illustrate it.

EXAMINATION E

Answer any ten.

1. (*a*) Graph and find the zeros of $y = x^2 - 5x + 4$.
   (*b*) Solve: 1) $x^3 + 8 = 0$, and 2) $3x^2 - 2x - 8 = 0$.

2. In the following, set down the determinant solutions for y. Evaluate the denominator only:

$$\begin{aligned} x - y + z + 3w &= 9 \\ 3x - y - z - w &= 7 \\ x + y + z + 5w &= 12 \\ x - y - 2z - 6w &= 0. \end{aligned}$$

3. Solve the following for x and y:

$$\begin{aligned} y &= 9 + 3x - 2x^2 \\ y + x - 3 &= 0. \end{aligned}$$

4. Write out the first, second, third, fourth, and eighth terms in the expansion of $\left(x - \frac{2}{x}\right)^{10}$.

5. (*a*) Find $\log \sqrt[3]{\frac{700 \times 0.002}{3^3}}$. (*b*) Find $\log_7 30$.
   (*c*) Find x if $70^x = 300$.

6. If $f(x) = \frac{1 + x}{1 - x}$, (*a*) find $f(\sqrt{2})$, (*b*) find $\frac{f(x) - f(-x)}{f(x) + f(-x)}$ and simplify.

7. A man has \$10,000 invested, part at 5% and the remainder at 6%. The interest for one year on the 5% investment exceeds the interest for one year on the 6% investment

by \$60. How much does he have invested at each rate? (Work algebraically.)

8, 9. Find all the roots of $2x^3 - 9x^2 + 8x + 5 = 0$. Check your result by graphing $y = 2x^3 - 9x^2 + 8x + 5$.

10. If five balls are selected at random from a bag containing 6 white, 4 black, and 7 red balls, what is the probability that the five will consist of two red and 3 white balls?

11. The following are reducible to quadratics by means of suitable substitutions.
Solve:

(*a*) $x^2 - 5x + 2\sqrt{x^2 - 5x + 10} = -2$

or (*b*) $\dfrac{x^2 + 1}{x} + \dfrac{x}{x^2 + 1} = \dfrac{5}{2}$.

12. Answer two of the following: (*a*) State and prove the remainder theorem. (*b*) Derive the formula for the sum and product of the roots in the equation $ax^2 + bx + c = 0$. (*c*) Prove $\log MN = \log M + \log N$.

EXAMINATION F

Answer any ten.

1. (*a*) If $y = f(x) = \dfrac{3x + 4}{4x - 3}$
find $f(y)$ in terms of x,
or (*b*) solve $\sqrt{2x - 1} - \sqrt{x + 3} = 1$.

2. Answer two parts:
(*a*) find $\log \sqrt[3]{\dfrac{3747(0.003)^2}{0.2}}$,
(*b*) find x if $375^x = 200^x\,(0.3)$,
(*c*) find $\log_{20}375$.

3. Do not evaluate the determinants in the following.
(*a*) Solve for y by determinants if $2x - 3y + 7z = 4$, $y - 3z + x + 1 = 0$, $z + 3x + 4y = 8$.
(*b*) Under what condition will $kx - 3y - 5 = 0$, $3x + y - 16 = 0$, $kx + 6y - 5 = 0$ be consistent?

4. Evaluate the following determinant: $\begin{vmatrix} 4 & 7 & 5 & -3 & 1 \\ 1 & 1 & 2 & 1 & 0 \\ 0 & 2 & 3 & -4 & 0 \\ 3 & 2 & 1 & 0 & 0 \\ 1 & -1 & 1 & 1 & 0 \end{vmatrix}$.

5. Find k so that $x^2 - 3kx + 4k + 1 = 0$ shall have:
(*a*) equal roots, (*b*) 3 for a root, (*c*) one root equal to 0,
(*d*) the product of its roots equal to 9, (*e*) the sum of its roots equal to 3.

6. Solve $y^2 + 3xy = 28$, $x^2 + y^2 = 20$.

7. Find the first four and the 7th term in the expansion of $(x^3 - 2y^2)^9$.

8. (*a*) Find the sum of all integers between 200 and 900 that are divisible by 7. (*b*) Find the limiting value of 0.6272727. . . .

9. State four theorems concerning the roots of a quadratic equation and prove one of them.

10. How many words can be formed from 8 consonants and 5 vowels if each word is to consist of 4 consonants and 2 vowels?

11, 12. Sketch $y = f(x) = 2x^3 - 5x^2 - 8x + 6$ and find all the roots of $f(x) = 0$.

13. Write the following in the form of an equation: the bend (B) of a rod supported at both ends varies directly as the weight (M) hung at its middle point, directly as the cube of the length (L) of the rod between supports, inversely as the width (W) of the rod, and inversely as the cube of the depth (D).

EXAMINATION G

Answer ten questions including number 10.

1. Factor each of the following:
(*a*) $x^4 - y^4 =$ (*b*) $x^{2n} - y^4 =$ (*c*) $x^2 - y^2 + 2y - 1 =$.

2. Simplify: (*a*) $\dfrac{\dfrac{a}{b^2} + \dfrac{b}{a^2}}{\dfrac{1}{a^2} - \dfrac{1}{ab} + \dfrac{1}{b^2}}$ (*b*) $1 - \dfrac{1}{1 - \dfrac{1}{x}}$.

3. Solve simultaneously $xy + 2x = 5$ and $2xy - y = 3$.

4. (*a*) Develop the formula for solving the equation

$$Ax^2 + Bx + C = 0.$$

(*b*) Write the first three and the 9th terms of $(x - 2y)^{57}$.

5. State three theorems concerning a quadratic equation in one unknown, and prove one of them.

Define: function, logarithm, discriminant of a quadratic.

6. Solve for "y" **only**, using determinants.

$$\begin{aligned} 2x - y + 3z &= 35 \\ x + 3y + 2z &= 15 \\ 3x + 4y &= 1. \end{aligned}$$

7. Find log $\dfrac{\sqrt{21}\cdot(24)}{1.5}$.

8. Simplify each of the following:

(*a*) $(10^{-1} - 10^{-2})^{1/2}$; (*b*) $(-\frac{1}{8})^{2/3}$; (*c*) $(a^{1/2} - b^{1/2})^2$.

9. (*a*) State and develop the formula for the sum of an arithmetic progression.

(*b*) Express the repeating decimal 0.181818 . . . as a fraction.

10. State as many facts as you can concerning the roots of $x^5 + 2x^4 - 7x^3 - 14x^2 - 18x - 36 = 0$. *Find all the roots.*

11. (a) In how many different orders can 6 people be seated at a round table?

(*b*) In how many ways can 6 people be arranged in a straight line?

(*c*) If $C(n, 5) = C(n, 4)$, find n.

12. Evaluate the determinant: $\begin{vmatrix} 1 & 1 & 2 & 1 \\ 0 & 2 & 3 & -4 \\ 3 & 2 & 1 & 0 \\ 1 & -1 & 1 & 1 \end{vmatrix}$.

EXAMINATION H

Answer ten questions.

1. Simplify: $\dfrac{(x - a) - \dfrac{x^2 + a^2}{x + a}}{(x + a) - \dfrac{(x - a)^2}{x + a}} = \cdot$

2. Evaluate: $\begin{vmatrix} 1 & -1 & 0 & 2 \\ 0 & 1 & 2 & -2 \\ 3 & 1 & 1 & 1 \\ -1 & 1 & 2 & 2 \end{vmatrix}.$

3. Solve for "z" only using determinants:

$$\begin{aligned} x + y + 2z &= 4 \\ x + 2y + z &= 4 \\ 2x + y + z &= 4. \end{aligned}$$

4. Define: logarithm, function, root of an equation, imaginary number.

   Show that if $f(x) = x^2 + x + 1$, then $f\left(\dfrac{-1 + \sqrt{-3}}{2}\right) = 0.$

5. (*a*) A can shovel the snow from a walk in 2 hours. B can do it in 3 hours. How long will it take both of them working together?

   (*b*) What is the probability of drawing one white and one black ball from a bag containing 5 white and 6 black balls, if two balls are drawn?

6. (*a*) Write the first four and the 14th term of $(\sqrt{2x} - y)^{15}$.

   (*b*) What fraction is represented by the repeating decimal .123123 . . . ?

7. Find log 15.

   Solve the equation $3^x = 200$. What is the log of 200 to the base 3?

8. (*a*) Develop the formula for solving $ax^2 + bx + c = 0$.

   (*b*) What is the discriminant of a quadratic equation? What does it tell us concerning the roots?

(*c*) Determine "k" so that the sum of the roots of $kx^2 + 7x + k = 0$ shall equal 14.

9. Solve simultaneously and graph: $y^2 = 8x$ and $2x + y = 8$.

10. (*a*) Rationalize $\frac{2\sqrt{3} - 1}{1 + \sqrt{3}}$. (*b*) Factor $x^6 - y^6$ into 4 factors.

    (*c*) Simplify $\left(\frac{2x^{-7}}{y}\right)^0 \div \frac{3}{x^{-2}}$.

11. Find all of the roots of: $x^4 + 2x^3 - 9x^2 - 10x - 24 = 0$.

12. State: the remainder theorem; Descartes' rule of signs; the fundamental theorem of algebra; a theorem concerning the number of roots of a polynomial. Prove one of them.

13. Write an equation whose roots are each three less than the roots of the equation given in problem 11.

    What facts can you state about a polynomial of odd degree whose signs are alternately + and −?

    What is the sum of the roots of the equation

$$x^3 + 5x - 2 = 0?$$

ANSWERS TO EXAMINATION QUESTIONS

Examination A.

1. See definition p. 50. $x = 0.$
2. See pp. 73-75.
3. (*a*) $x-1$, (*b*) $\frac{4i-7}{13}$, (*c*) $A - \frac{1}{A}$.
4. $(5, 0)$, $(0, 5)$.
5. (*a*) $k = \frac{4 \pm 2\sqrt{22}}{9}$, (*b*) $k = -4$,

   (*c*) $k = -1$, (*d*) $k = 17$.
6. (*a*) 1.1367, (*b*) $x = 1.771$.
7. $$(2x)^{21} + 21(2x)^{20}(-y) + \frac{21 \cdot 20}{2!}(2x)^{19}(-y)^2 + \frac{21 \cdot 20 \cdot 19}{3!}(2x)^{18}(-y)^3 + \frac{21 \cdot 20 \cdot 19 \cdot 18}{4!}(2x)^{17}(-y)^4 + \cdots$$

   Eleventh term is $\frac{21 \cdot 20 \cdot 19 \cdots 14 \cdot 13 \cdot 12}{10!}(2x)^{11}(-y)^{10}$.
8. 5.
9. See pp. 151-153.
10. See pp. 166, 158, 160, 159.
11. (*A*) See p. 162,

    (*B*) $C(13, 10) = C(13, 3) = \frac{13 \cdot 12 \cdot 11}{3!} = 286.$
12. $-1, -2, 1, \frac{-1+i\sqrt{3}}{2}, \frac{-1-i\sqrt{3}}{2}$.
13. $$y = \frac{\begin{vmatrix} 1 & 0 & 2 \\ 2 & -1 & -2 \\ 3 & -1 & 6 \end{vmatrix}}{\begin{vmatrix} 1 & 1 & 2 \\ 2 & -1 & -2 \\ 3 & -5 & 6 \end{vmatrix}} = \frac{1}{8}.$$

Examination B.

1. (*a*) $\frac{5}{4}$, (*b*) $\pm\frac{1}{8}$.
2. (*A*) $f(3) = 5$, $f(-2) = 5$.
   (*B*) Area $= \frac{1}{2} h (h + 1)$.
3. (*a*) $\left(\frac{1+\sqrt{17}}{2}, \frac{1-\sqrt{17}}{2}\right)$, $\left(\frac{1-\sqrt{17}}{2}, \frac{1+\sqrt{17}}{2}\right)$.
4. *A*. (*a*) 1.0791, (*b*) 0.1276, (*c*) 7.1761 − 10,
   (*d*) 18.1636.

   B. $x = \frac{1.2552}{1.0791} = 1.163$.
5. $-\frac{1}{2}$, $\frac{-1+i\sqrt{3}}{2}$, $\frac{-1-i\sqrt{3}}{2}$.
6. See p. 158.
7. (*A*) See p. 140. (*B*) Sum = ⅓.
8. Width = 3 in.
9. $y = -1$.
10. (*A*) See p. 148. (*B*) $32x^5 - 80x^4 + 80x^3 - 40x^2 + 10x - 1$.
11. (*A*) $z = 1$. (*B*) First part, $k = \pm 4$; Second part, $k = -13/3$.
12. (*A*) 10,080.
    (*B*) (*a*) $\frac{5}{14}$, (*b*) $\frac{15}{28}$, (*c*) $\frac{3}{28}$.

Examination C.

1. See p. 50.
2. (*a*) See p. 73.
   (*b, 1*) $k = 4$, (*b, 2*) $k = -2$.
3. (2, 4), (1, 3).
4. (*a*) $x^3 - 5x^2 + 11x - 15 = 0$.
   (*b*) $x = 1$.
5. (*a*) $2\frac{10}{27}$, (*b*) $92{,}378x^{10}y^{9/2}$.
6. (*a*) 0.5836, (*b*) 1.771, (*c*) see p. 122.

7. (*a*) See p. 158.
   (*b*) $-1, 2, 1 + \sqrt{3}, 1 - \sqrt{3}$.

8. (*a*) 4, 6, 8, 10.
   (*b*) $l = 486$, $s = 728$.

9. $\frac{175}{572}$.

10. (*a*) $x = \frac{\begin{vmatrix} 5 & -1 & 2 & 0 \\ -7 & 1 & -2 & -3 \\ 3 & 1 & 1 & 3 \\ 7 & -2 & 0 & 7 \end{vmatrix}}{\begin{vmatrix} 2 & -1 & 2 & 0 \\ 0 & 1 & -2 & -3 \\ 1 & 1 & 1 & 3 \\ 1 & -2 & 0 & 7 \end{vmatrix}}$, (*b*) 75.

11. $s = \frac{64}{7}$.

12. (*a*) $4 + 9i$, (*b*) $1 + 2i$.

Examination D.

1. $-3 + 5i, -3 - 5i$. $\frac{6}{25} - \frac{17}{25}i$.

2. $x = 3$ or -2. $x = 5$.

3. (*a*) $(1, -3), (3, 1)$.
   (*b*) $(\sqrt{7} + i, \sqrt{7} - i), (-\sqrt{7} - i, -\sqrt{7} + i)$.

4. 0.9360. See p. 120.

5. $I = 10.8$.

6. See pp. 138-141. Sum = 2450.

7. See pp. 141, 139, 72, 73.

8. $(2x)^{51} - 51(2x)^{50}(3) + \frac{51 \cdot 50}{2!}(2x)^{49}(3)^2 - \frac{51 \cdot 50 \cdot 49}{3!}(2x)^{48}(3)^3 + \cdots$ See pp. 198-200.

9. $z = \frac{129}{-129} = -1$.

10. $-2, -1, 1, \frac{-1 + i\sqrt{3}}{2}, \frac{-1 - i\sqrt{3}}{2}.$

$x^5 + 13x^4 + 66x^3 + 163x^2 + 193x + 84 = 0.$

$x^5 + 6x^4 + 8x^3 - 8x^2 - 48x - 64 = 0.$

11. (*a*) 6840, (*b*) 200.

12. See pp. 166, 160, 158

Examination E.

1. (*a*) $x = 1$ or 4.
   (*b*,*1*) $-2, 1 + i\sqrt{3}, 1 - i\sqrt{3}.$ (*b*, *2*) $2, -4/3.$

2. $y = \frac{\begin{vmatrix} 1 & 9 & 1 & 3 \\ 3 & 7 & -1 & -1 \\ 1 & 12 & 1 & 5 \\ 1 & 0 & -2 & -6 \end{vmatrix}}{\begin{vmatrix} 1 & -1 & 1 & 3 \\ 3 & -1 & -1 & -1 \\ 1 & 1 & 1 & 5 \\ 1 & -1 & -2 & -6 \end{vmatrix}}$ Denominator = 0.

3. $(-1, 4), (3, 0).$

4. $x^{10} - 20x^8 + 180x^6 - 960x^4$ - -, 8th term $= \frac{-15360}{x^4}$

5. (*a*) 9.5716 −10, (*b*) 1.748, (*c*) $x = 1.342.$

6. (*a*) $-(3 + 2\sqrt{2})$, (*b*) $\frac{2x}{1 + x^2}$

7. \$6,000 at 5%, \$4,000 at 6%.

8, 9. $\frac{5}{2}, 1 + \sqrt{2}, 1 - \sqrt{2}.$

10. $\frac{C(7, 2) \cdot C(6, 3)}{C(17, 5)} = \frac{15}{221}.$

11. (*a*) $x = 2$ or 3.
    (*b*) $1, \frac{1 + i\sqrt{15}}{4}, \frac{1 - i\sqrt{15}}{4}.$

12. See pp. 158, 73-74, 122.

Examination F.

1. (*a*) f(y) = x, (*b*) x = 13.

2. (*a*) 9.7423 − 10, (*b*) —1.915, (*c*) 1.974.

3. (*a*) $y = \frac{\begin{vmatrix} 2 & 4 & 7 \\ 1 & -1 & -3 \\ 3 & 8 & 1 \end{vmatrix}}{\begin{vmatrix} 2 & -3 & 7 \\ 1 & 1 & -3 \\ 3 & 4 & 1 \end{vmatrix}}$, (*b*) $\begin{vmatrix} k & -3 & -5 \\ 3 & 1 & -16 \\ k & 6 & -5 \end{vmatrix} = 0$, or $k = {}^{15}\!/_{16}$.

4. 48

5. (*a*) 2 or $-\frac{2}{9}$, (*b*) 2, (*c*) $-\frac{1}{4}$, (*d*) 2, (*e*) 1.

6. $(4, 2), (-4, -2), \left(\frac{\sqrt{10}}{5}, \frac{7\sqrt{10}}{5}\right), \left(\frac{-\sqrt{10}}{5}, \frac{-7\sqrt{10}}{5}\right)$.

7. $x^{27} - 18x^{24}y^2 + 144x^{21}y^4 - 672\,x^{18}y^6$ - - - -
   7th term = $5376\,x^9y^{12}$.

8. (*a*) 54,950. (*b*) $\frac{69}{110}$.

9. See pp. 73-75.

10. 504,000.

11, 12. $-\frac{3}{2}, 2 + \sqrt{2}, 2 - \sqrt{2}$.

13. $B = \frac{k\,M\,L^3}{W\,D^3}$.

Examination G.

1. (*a*) $(x - y)(x + y)(x^2 + y^2)$.
   (*b*) $(x^n - y^2)(x^n + y^2)$.
   (*c*) $(x - y + 1)(x + y - 1)$.

2. (*a*) $(a + b)$.
   (*b*) $\frac{1}{1 - x}$.

3. (1, 3), (5/4, 2).

4. (*a*) See Sec. 67, part d.

   (*b*) $x^{57} - 57x^{56}(2y) + \frac{57 \cdot 56}{2\,!}x^{55}(2y)^2.$

   Ninth term $= + C(57, 8)x^{49}(2y)^8.$

5. See Secs. 52, 68, 101.

6. $y = -2.$

7. 1.8651.

8. (*a*) 3/10, (*b*) $\frac{1}{4}$, (*c*) $a - 2\sqrt{ab} + b.$

9. (*a*) $s = \frac{n}{2}(a + 1),$ (*b*) 2/11.

10. $x = -2, 3, -3, -\sqrt{-2}, +\sqrt{-2}.$

11. (*a*) 120, (*b*) 720, (*c*) $n = 9.$

12. 48.

Examination H.

1. $-\frac{a}{2x}.$

2. $-34.$

3. $z = 1.$

4. Answer given in problem.

5. (*a*) 6/5 hours. (*b*) $p = 6/11.$

6. (*a*) $(\sqrt{2x})^{15} - 15(\sqrt{2x})^{14}y + \frac{15 \cdot 14}{2\,!}(\sqrt{2x})^{13}y^2$

   $- \frac{15 \cdot 14 \cdot 13}{3\,!}(\sqrt{2x})^{12}y^3 + \ldots.$

   Fourteenth term $= -C(15,13)(\sqrt{2x})^2y^{13} = -C(15,2)(2x)y^{13}.$

   (*b*) $\frac{41}{333}.$

7. $\log 15 = 1.1761$ $x = 4.82$ $\log_3 200 = 4.82.$

8. (*a*) See Sec. 67, part d.

   (*b*) See Sec. 68, item 3.

   (*c*) $k = -\frac{1}{2}.$

9. (2, 4) and (8, −8).

10. (*a*) $\frac{7 - 3\sqrt{3}}{2}$.

(*b*) $(x - y)(x + y)(x^2 + xy + y^2)(x^2 - xy + y^2)$.

(*c*) $\frac{1}{3x^2}$.

11. $x = 3, -4, \frac{-1 + \sqrt{-7}}{2}, \frac{-1 - \sqrt{-7}}{2}$.

12. See Chapter XVIII.

13. (*a*) $x^4 + 14x^3 + 63x^2 + 98x = 0$.

(*b*) At least one positive root; no negative roots.

(*c*) Sum = 0.

INDEX